TESTING OF METALLIC MATERIALS

Second Edition

TESTING OF METALLIC MATERIALS

Second Edition

Prof. A.V.K. Suryanarayana
Formerly Head, Metallurgical Section,
V.I.S.S.J. Polytechnic,
Bhadravathi.

BSP **BS Publications**
A unit of **BSP Books Pvt., Ltd.**

4-4-309/316, Giriraj Lane, Sultan Bazar,
Hyderabad - 500 095
Phone : 040 - 23445605, 23445688

Published by :

BSP BS Publications

A unit of **BSP Books Pvt., Ltd.**

4-4-309/316, Giriraj Lane, Sultan Bazar,
Hyderabad - 500 095
Phone : 040 - 23445605, 23445688
e-mail : info@bspbooks.net

ISBN : 978-93-52300-37-2 (HB)

Dedicated to

my wife

Smt. Annapurna

Preface to Second Edition

The essence of this new edition is to provide an insight into the methods of testing, the view point. The chapter dealing with fracture mechanism is elaborated to give a better perspective of failure from a physical metallurgical concept. Elementary ideas regarding grain structure and critically resolved shear stress of single crystals under load are provided. It is hoped that this will further enhance the value of this book. I expect the same patronage for this edition from the engineering fraternity.

- Author

Preface to First Edition

Testing of metallic materials is a common engineering subject. It deals with the various testing techniques and calculation of the different strengths, etc., and values. With the advances made in the field of Materials Science, present-day testing has changed its character. From the old method of reporting test results, the interpretation of the results of tests has taken over a prominent position. The metallurgist enters into the picture here.

For the last so many years, since I was a student of Metallurgical Engineering, the need for a comprehensive textbook has been felt. This need has remained unfulfilled; hence, I have undertaken the venture of providing a book for students of engineering in general, and metallurgy in particular.

I have tried my best to present the subject in such a way that the book will be useful for self-study. Pure science graduates who are working in the industry and engaged in testing are also kept in view while treating the subject. The necessary metallurgical interpretations are given briefly, at a fundamental concept level.

This branch of engineering is such that not much research has taken place, though ductility has attracted the attention of many researchers on the Continent. Their work is not touched upon, firstly due to its advanced and specialized nature, and secondly the lack of much information. Similarly, details regarding loading devices, load measuring and strain measuring devices are not touched upon wantonly. Yet, I think it will not be a drawback to the value of this volume as a whole.

Various organizations, mostly from the United States, United Kingdom and also from West Germany and Belgium, have readily responded to my requests for their machinery, photographs and other data. They are individually acknowledged at the appropriate places in the book. I also express my gratitude to Mr. A. G. Fenner, John Adam Street, London, for his kind consent in allowing me to use some of his material. I should further like to express my gratitude to Dr. R. E. Reed-Hill of the University of Florida, U.S.A., and Dr. C. Suryanarayana of the Banaras Hindu University, Varanasi, for their reviews and suggestions regarding the manuscript.

If the student community and the industrial workers receive this book with a sense of appreciation, I will consider my ambition fulfilled. I appeal to the teaching fraternity to kindly let me know of any omissions and commissions with a view to rectifying them in future. Suggestions, from all quarters, are welcome.

- Author

CONTENTS

Introduction

There are millions of materials and their uses are also multitudinous. This large number of meterials came into existence because of the various number of industrial needs for them. The particular property or combination of properties a material should possess, depends on its specific use and the service conditions.

With the development of various machines, the structures of most of them far from simple, the necessity for materials which possess properties within narrow specified ranges has arisen. The material properties can be classified into three major headings:

(i) Physical,

(ii) Chemical, and

(iii) Mechanical.

Physical Properties

Density or specific gravity, porosity, moisture content, etc., can be classified under this category.

Chemical Properties

Many chemical properties come under this category. These include acidity or alkalinity, reactivity and corrosion. The most important of these is corrosion which can be explained in layman's terms as the resistance of the material to decay while in continuous use in a particular atmosphere.

Mechanical Properties

Mechanical properties include the strength properties like tensile, compression, shear, torsion, impact, fatigue and creep. The tensile strength of a material is obtained by dividing the maximum load which the specimen bears by the area of cross-section of the specimen. In the tensile test, the material is loaded in such a way that tensile forces are brought to play in it. The test specimen is fixed at one end and loaded at the other, the load acting as if it is elongating the specimen.

Exactly opposite to tension type of loading occurs in compression testing. The two loads are axial and opposite to each other.

In shear, the conditions of compression prevail except that the two opposing loads are not in the same axis.

The torsion test consists of twisting the body of a specimen over its length, and fixing it at one end.

In the impact test, a sudden or shock load is used to bring about the failure of the material and the energy absorbed is noted.

Fatigue is a long-time test in which the specimen is subjected to repeated loads of a small magnitude, whereas the creep test can be practically considered as a high temperature tensile test.

Besides these, a very common test is that of hardness. Various types of hardness tests are in vogue depending upon the nature of hardness to be determined.

Taking into account all these, we can classify the tests into two major categories as follows :

1. Tests (or reactive tests), and
2. Service condition tests.

All the tests except impact, creep and fatigue can be classed as reactive tests. Fatigue, creep and impact tests represent, to a large extent, the actual service conditions of the object under consideration.

In any testing process, the actual use of the material is kept in view and the necessity for the test is first decided. If a material is to be used for the manufacture of a lathe chuck, its hardness as well as compression and shear strengths should be appreciably high. If a material is to be used for a bearing bush, however, its porosity is the major consideration, so as to facilitate a liberal oil absorption; it should also be relatively soft compared to the shaft or spindle on which it is to work, because it would be very expensive to replace a shaft itself, if it is worn out. On the other hand, if the bush is worn out quickly, it is easier and cheaper to replace. A chisel steel should be hard and possess a high impact strength, and so on.

Testing as it is, is an innate human tendency. A layman who purchases a glass tumbler, rings it with his fingernail. Of course, he does so with the knowledge that it reveals any internal defects or cracks in the form of a dull sound. It is a human tendency not to be a loser in a bargain. When we pay for something, we wish to purchase the best for that price. This is one aspect of testing.

Next, when we wish to order a particular object or commodity, we should describe it in full, so as to enable the supplier to understand our need. A description to the size and shape would do if it is an ordinary object like paper, a pencil, or such. But in the case of an engineering object, a simple description will not do. The object will serve us properly only when it suits the purpose from all standpoints. In such a case, the type of loads it experiences in service and their magnitude are to be taken into account and minimum values for the same should be specified, besides its dimensions. Such a description of the object is termed as its specification.

Thus, after realizing the need for a specification, attempts were made to perfectly and comprehensively specify the various characteristics and properties. Mass production, which is the highlight of this century, has made it all the more necessary to cut down the number of sizes and shapes for the same products, so as to facilitate inter-changeability. Delinquency on the part of greedy suppliers which had become the bane of engineering society, has further necessitated rigid stipulations regarding all the expressable characteristics and the properties of the materials.

This has given a fillip to the standardization movement. Various standards organizations such as the American Society for Testing and Materials (ASTM), the Society of Automotive Engineers (SAE), American Foundrymen's Society (AFS), the British Standards Institution (BSI), and so on, have standardized and are still standardizing various products. In India, the Indian Standards Institution was founded after Independence. It is, now known as Bureau of Indian Standards, responsible for pioneering work in the field of standardization. Besides, they have introduced the certification marks scheme for various products. Under this, manufacturing organizations who oblige and conduct the prescribed testing of the product according to the relevant Indian standard, will be licensed to put an I.S.I. mark on their products. The existence of the mark itself is a guarantee to the consumer that the product conforms to the relevant Indian standard. Similar practices are in vogue in some other countries like France.

Going back to the testing, we have underlined the need for a test in the foregoing paragraphs. A majority of the mechanical tests are of a destructive type, i.e., the specimen tested is either broken or becomes unusable after the test. Methods of non-destructive testing* are also available but their application is limited.

While we can appreciate the importance and significance of materials testing, we will also realize that if we go on testing every product manufactured (and thereby breaking it) there will be nothing left to market. We cannot afford to do so. A hundred per cent check may be made for dimensions, or some other property that may be revealed by a 'nondestructive' method, but it will be insensible to conduct tensile or fatigue tests on a hundred per cent of the production. The ready solution to this problem is the testing of representative samples. In this, a sample which represents a batch of the product, is taken and tested. If the product manufactured is like a powder, a representative sample is taken by the process of a 9-point circle. If the product is of the texture of crushed ore, 'coning and quartering' is adopted. In the case of engineering items, samples are taken at a prescribed frequency during manufacture. That is, a sample in the first hour of production, another in the next hour and so on. There are other methods also as prescribed in the statistical quality control methods**. Whatever may be the method prescribed, a deep insight into the method of sampling reveals that a maximum representation of the lot is obtained in the sample.

Next to taking the sample, comes the testing proper. Various methods of testing are discussed in the ensuing chapters. When we test a piece of wood of the order of the trunk of a small tree (approximately 15 cm diam.) we record a particular load when it breaks. If we test a piece of copper in the form of a thick wire, we may get the same value as in the previous case. However, in spite of having obtained the same breaking load values for the two materials, we cannot say that both the materials are equally strong. Again, if we take two pieces of the same metal, one double the other in size, and test them, we record differing loads of breakage for the two. This type of testing leads us nowhere, for when we perform a test, it should be capable of (i) being repeated, and representative of the material concerned, and (ii) serving as a standard or basis for comparison. For this reason, we should take a specimen. The specimen should be of a definite size and shape so that when we test the other materials also in the same form, identical conditions of testing are maintained. Then alone can we obtain results worth comparison and classification.

Standard test specimens for different forms of materials, such as those in the form of wire, plate, tubing, and solids, specified for each type of test are described in the relevant chapters.

---

* Refer to the latter half of the book.

** Refer I.S.12 of 1960.

PART - I

CHAPTER 1

Tensile and Torsion Tests

TENSILE TEST

The strength of a material when it is called upon to withstand loads which produce a tensile stress in it, is defined as the tensile strength of the material. A tensile test is the first of its kind of tests. In routine usage, the term "tensile" is omitted and tensile strength is referred to simply as strength.

The conditions of tension are too common to be described in detail. A load which tends to pull apart the two ends of an object is said to be a tensile load. Similarly, readers should be familiar with the terms compression, shear and torsion from earlier education. However, these are illustrated schematically in Figs. 1.1 a to 1.1 d.

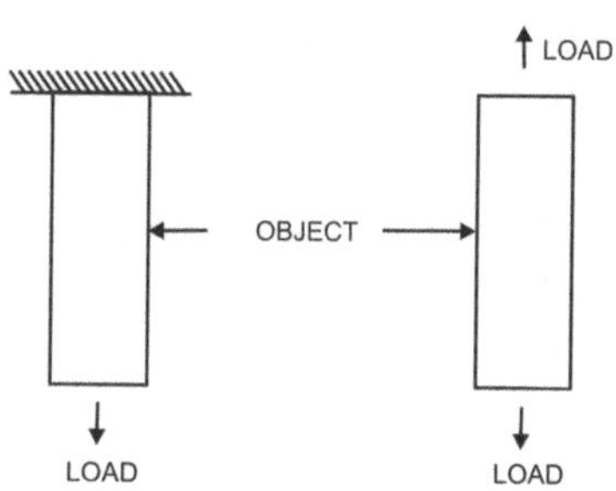

Fig. 1.1a. Tesion.

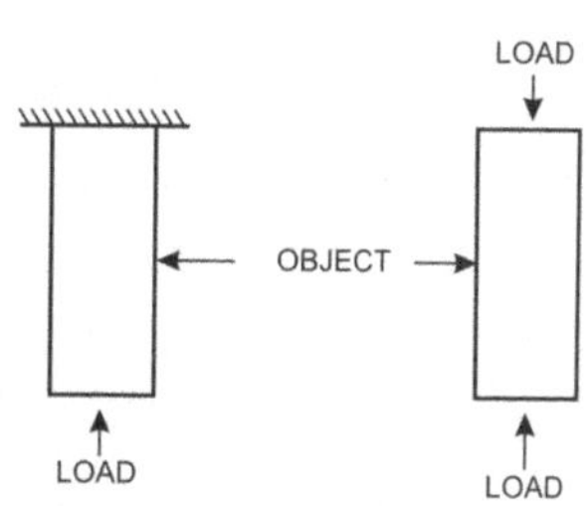

Fig. 1.1b. Compression.

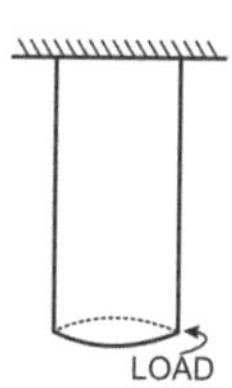

Fig. 1.1c. Torsion.

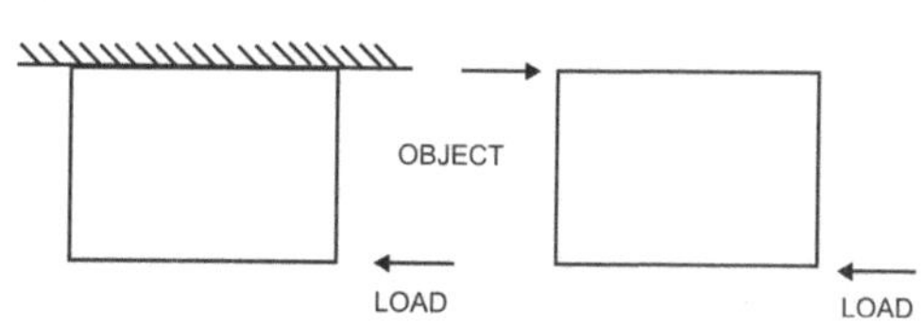

Fig. 1.1d. Shear.

Before proceeding to the testing it is appropriate to recapitulate some elementary terminology such as load, stress, deformation, strain, etc.

Load

From our standpoint, load is referred to as the weight applied to the body in testing. If a wire is suspended from a nail on a wall and a kilogram-weight is placed in a pan fixed to its bottom end, it is said that the wire is loaded. In this example, the magnitude of the load applied is one kilogram.

In any test, load and its application are important. Load is applied to the extent desired, viz., to the extent the material under test can withstand it before it breaks. The method of application of the load varies depending upon the test performed. In tensile testing the load applied is tensile in nature, i.e., it tends to pull or elongate the specimen. In compression testing, the specimen is subjected to a compressive load, i.e., it is compressed between two opposing loads, and so on. The units for the load are force units and the loads are expressed in weight-Kgf.

Stress

When a body is loaded, forces will be set up inside it. These forces will be in a direction opposite to that of the application of the external load. The magnitude of these internal forces will be such that the effect of the external load is balanced and the body remains in mechanical equilibrium. These internal forces are called stresses. Stress is defined as the internal reaction set up in the solid per unit cross-sectional area. It should be further noted that the internal reaction is proportional to the load applied. At any load, the value of the stress in the solid can be arrived at by dividing the load by the cross-sectional area over which it acts. Stress is express in Kgf/mm^2.

Deformation

When a load is applied on a solid, it tends to change the shape of the solid in its direction. The change affected is called deformation. In the case of tensile loading, the deformation suffered by the object is termed as extension or elongation. Deformation is expressed in the same units as the particular dimension is expressed, i.e., in mm or cm.

Strain

We have seen above that the deformation is the total change suffered by the object in the particular dimension. Strain is the deformation expressed on the basis of the unit dimension. It is customary to express the deformation as strain because the term strain gives a precise measure. In the example illustrated (Fig. 1.2), the strain is l/L. The units for strain are mm per mm. Thus, it is a number.

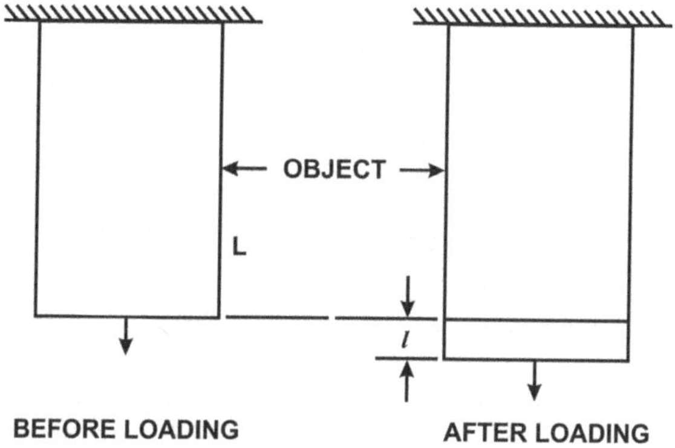

Fig. 1.2a. Deformation.

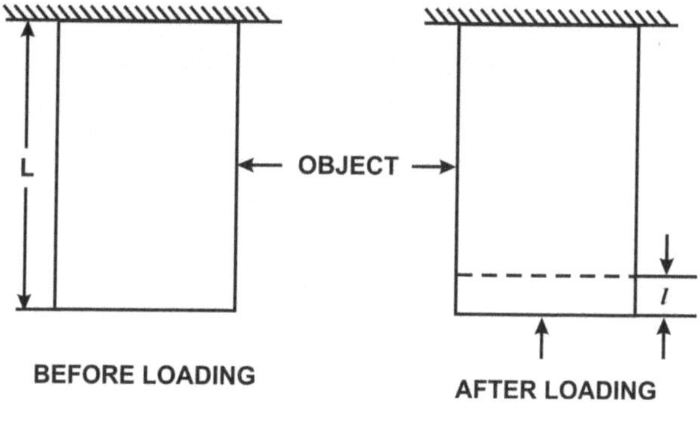

Fig. 1.2b. Compression.

Stress-Strain Diagram

This is a curve plotted between the stress along the *Y*-axis (ordinate) and the strain along the *X*-axis (abscissa) in a tensile test. A material tends to change or changes its dimensions when it is loaded, depending upon the magnitude of the load. When the load is removed it can be seen that the deformation disappears. For many materials this occurs up to a certain value of the stress called the elastic limit. This is depicted by the straight line relationship and a small deviation thereafter, in the stress-strain curve (Fig. 1.3).

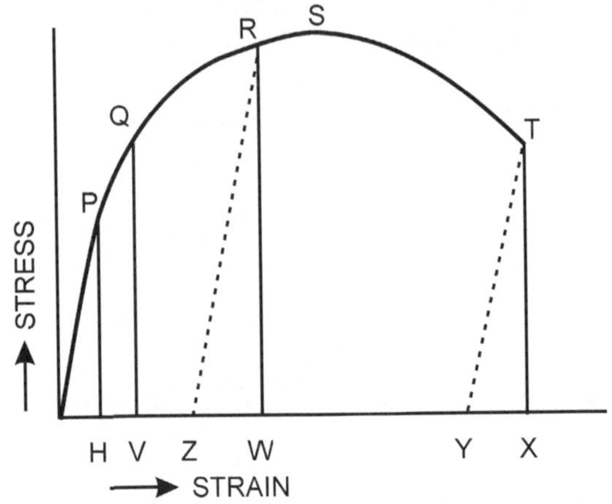

Fig. 1.3 Stress-strain curve.

Within the elastic range, the limiting value of the stress up to which the stress and strain are proportional, is called the limit of proportionality. In this region, the metal obeys Hooke's law, which states that the stress is proportional to strain in the elastic range of loading (the material completely regains its original dimensions after the load is removed). In the actual plotting of the curve, the proportionality limit is obtained at a slightly lower value of the load than the elastic limit. This may be attributed to the timelag in the regaining of the original dimensions of the material. This effect is very frequently noticed in some non-ferrous metals.

While iron and nickel exhibit clear ranges of elasticity, copper, zinc, tin, etc., are found to be imperfectly elastic even at relatively low values of stresses. Actually the elastic limit is distinguishable from the proportionality limit more clearly depending upon the sensitivity of the measuring instrument.

When the load is increased beyond the elastic limit, plastic deformation starts. Simultaneously the specimen gets work-hardened.

A point is reached when the deformation starts to occur more rapidly than the increasing load. This point is called the yield point *Q*. The metal which was resisting the load till then, starts to deform somewhat rapidly, i.e., yield.

The elongation of the specimen continues from *Q* to *S* and then to *T*. The stress-strain relation in this plastic flow period is indicated by the portion *QRST* of the curve. At *T* the specimen breaks, and this load is called the breaking load. The value of the maximum load S divided by the original cross-sectional area of the specimen is referred to as the ultimate tensile strength of the metal or simply the tensile strength.

Logically speaking, once the elastic limit is exceeded, the metal should start to yield, and finally break, without any increase in the value of stress. But the curve records an increased stress even after the elastic limit is exceeded. Two reasons can be given for this behaviour:

(i) the strain hardening of the material, and

(ii) the diminishing cross-sectional area of the specimen, suffered on account of the plastic deformation.

The more plastic deformation the metal undergoes, the harder it becomes, due to work-hardening. The more the metal gets elongated the more its diameter (and hence, cross-sectional area) is decreased. This continues until the point *S* is reached.

After *S*, the rate at which the reduction in area takes place, exceeds the rate at which the stress increases. Strain becomes so high that the reduction in area begins to produce a localized effect at some point. This is called necking.

Reduction in cross-sectional area takes place very rapidly; so rapidly that the load value actually drops. This is indicated by *ST*. Failure occurs at this point *T*.

Once the fracture occurs, the load is removed from the specimen. The elastic part of the deformation, i.e., *XY* is immediately recovered by the specimen. Thus it should be noted that whatever elastic deformation was there in the specimen, it is recovered after the fracture.

We have seen above that once the plastic deformation has started, the cross-sectional area of the specimen starts to decrease. Thus, the actual stress at any moment after the plastic deformation has started is much more. This is called the true stress. Sometimes a curve is plotted between the true stress and true strain. In such a curve the stress value rises and does not show any dipping as in the case of the conventional stress-strain curve shown by the dotted curve in Fig. 1.4a.

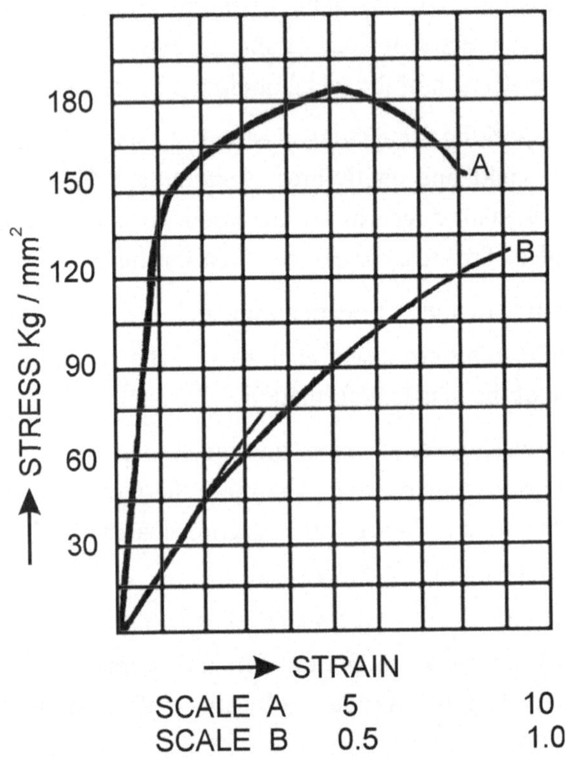

A : Engineering stress-strain curve B : True stress-strain curve

Fig. 1.4 a. Hardened Ni-Cr steel.

Yield Point

Consider the stress-strain diagram of mild steel (Fig. 1.4b). The material has a high proportional limit. When the load is increased beyond the proportional limit, a point is reached when the specimen suddenly starts to deform at a faster rate without any increase in the load. The highest value of stress after which this sudden extension occurs is known as the upper yield point (*Yu*). The lower yield point (*Yl*) is the stress which produces a considerable extent of elongation. The upper yield point is depend upon the size and shape of the specimen, its surface finish, and the rate of loading. In routine testing it is the lower yield point which is measured.

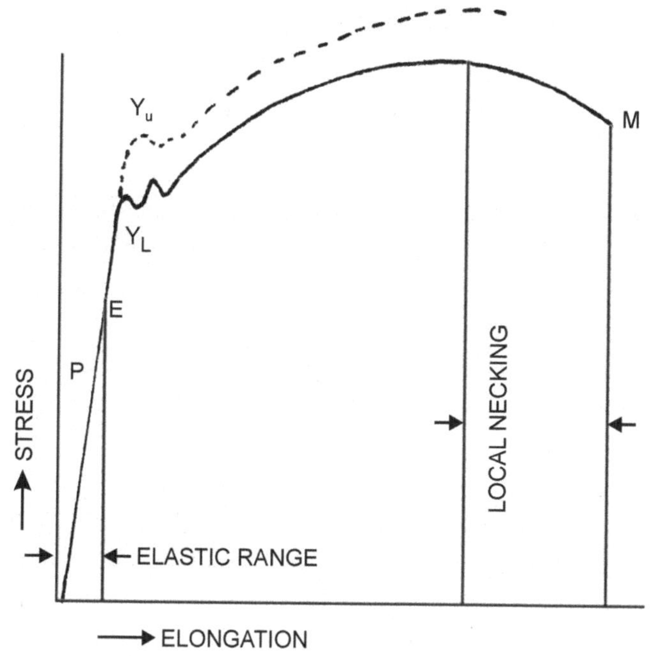

Fig. 1.4 b. Stress-strain curve of mild steel.

The deformation at the yield point is only local in nature. It starts at one point and that region gets work-hardened; so the flow starts again at a region adjacent to the former region. Hardening occurs here too, and the process continues. Thus, the flow is spread throughout the specimen. Each successive work-hardening tends to increase the stress. But the effect is only momentary, and again the stress value falls, due to the flow in the neighbouring region. This is the reason why ups and downs are noticed on the curve in the yield point region (Lüders bands).

As a result, the entire specimen gets work-hardened and the stress begins to rise. Deformation becomes uniform.

Yield Point Phenomenon

It is established that the yield point phenomenon is exhibited by metals and alloys of the body-centred-cubic and hexagonal-close-packed crystal structures which form interstitial solid solutions. Only the four elements carbon, oxygen, nitrogen, and boron form interstitial solid solutions. Thus steel is the most common alloy which exhibits this property. This phenomenon is best explained by the edge dislocation movement. The edge dislocation consists of the compression and tension portions in the slipped and unslipped regions respectively. When it is viewed isometrically, it looks like a pipe pierced through the metal lattice. Calculations show that the diameter of the pipe is much larger than the diameter of the interstitial hole. As a result, the interstitial atoms segregate along the dislocation lines for, in such a configuration, the 'energy' value is the lowest. It is further shown that the screw dislocations also react with the solute atoms. This interaction is very strong when the lattice is non-symmetrically deformed so that a tensile component of the stress is developed. This is called the anchoring of the dislocation. An anchored dislocation will not move and resists the stress. It moves only if a higher stress, sufficient to overcome this anchoring effect of the solute atoms, is applied. When the value of stress reaches this critical value, the anchored dislocations are torn away all of a sudden and rendered free. This is evident by the appearance of the first Lüders band. This triggers the operations in other planes at various other sources. Continued deformation will be evident at a slightly smaller stress, called the lower yield point (*Yl*). These manifest as the ups and downs in the stress-strain curve and as Lüders bands on the specimen.

It should be noted that the yield point phenomenon is not noticeable above a critical temperature, for the concentration of the solute atoms at the dislocation sites will not be much to make the anchoring effect strong.

Percentage Elongation and Reduction in Area

Before the test is made, the gauge length is marked on the standard specimen (L_o). After the specimen is broken, the two pieces are kept together as if the specimen is not broken at all, with the two fractured surfaces matching each other (Fig. 1.5). The distance between the two

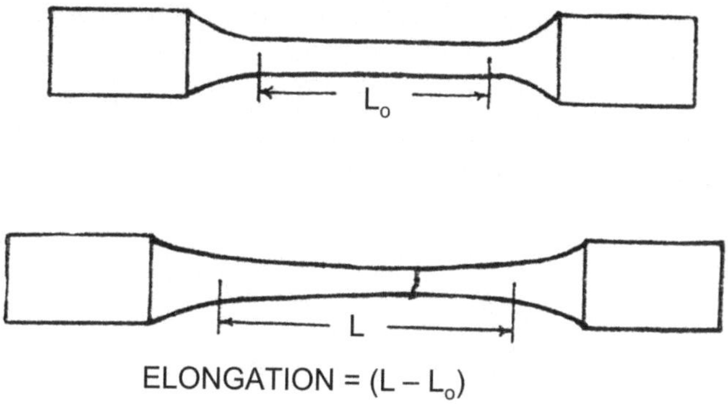

Fig. 1.5 Percentage elongation.

gauge length marks is again measured (L). The elongation, and therefrom the percentage elongation, are computed as follows :

Gauge length	$= L_o$
Distance between the gauge length marks after the failure	$= L$
Elongation	$= (L - L_o)$
Percentage elongation	$= \dfrac{(L - L_o)}{L_o} \times 100.$

Elongation suffered by the specimen owing to the application of the load can be classified into categories, as follows :

(i) The uniform elongation which has occurred until the maximum value of the load S is reached, and

(ii) the local extension produced in the specimen on account of necking. This has occurred after the maximum load S is exceeded.

The uniform elongation is dependent upon the gauge length. As the length of the specimen is increased, the effect of this factor on the total elongation increases. The local extension produced, however, is independent of the gauge length, but it varies with the area of cross-section of the specimen. Thus, it can be said that if the gauge length of the specimen is increased, the effect of necking on the elongation value is decreased*.

* It is recommended to use specimens of the same gauge length while studying the ductility behaviours of different metals, from the point of view of the percentage elongation.

This relationship is graphically shown in Fig. 1.6. This is the reason why the gauge length of the specimen is always mentioned while reporting or specifying the percentage elongation. If the failure occurs outside the gauge length marks, the correct value of the elongation cannot be as certained. It is recommended to ignore the test and repeat the test using another specimen.

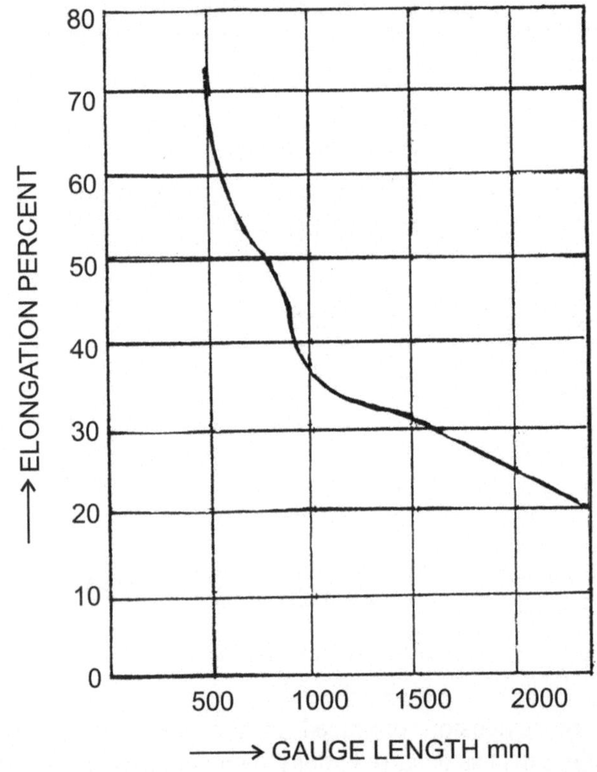

Fig. 1.6 Variation of percentage elongation depending upon gauge length in a low carbon steel.

The test results would be comparable only when the test specimens are geometrically similar, i.e., the ratio of the length (gauge length) to the square root of its diameter is maintained constant (Barba's formula). In India and the United Kingdom the value $L/\sqrt{D}$ is specified as 5.65 as per the recent I.S.O. recommendation.

Percentage reduction in area is the decrease in the cross-sectional area of the specimen up to failure, expressed as a percentage of the original cross-sectional area. Though both percentage reduction in area and elongation indicate the ductility of the material, the former is, in particular, a guide to the formability behaviour of the metal. Though Hadfield manganese steel, stainless steel, copper and aluminium exhibit high values of percentage elongation, the former two possess lower percentage reduction in area. Consequently, they work-harden rapidly and so are not suitable for cold working as aluminium and copper are.

Proof Stress

In many materials including high-carbon and alloy steels and nonferrous alloys, the stress-strain diagram does not indicate a well-defined straight line portion and yield point (Fig. 1.7). Thus, it is not possible to obtain the elastic limit. For engineering design the yielding nature of the material is important. Hence, in the case of such materials a "proof-stress" at a specified strain is calculated. The strain value for the calculation of proof-stress is specified in terms of the gauge length (as a small percentage). Thus, the stress corresponding to a certain allowable amount of plastic deformation is measured and taken in place of the proportionality limit. In checking for acceptance, the material is loaded with the load corresponding to the proof-stress values for 15 secs and the load removed. The material is deemed to have passed the test if it does not show a permanent set, greater than the specified precentage of the gauge length called offset. Usually o.1% permanent set is prescribed for evaluating proof-stress (Fig. 1.8).

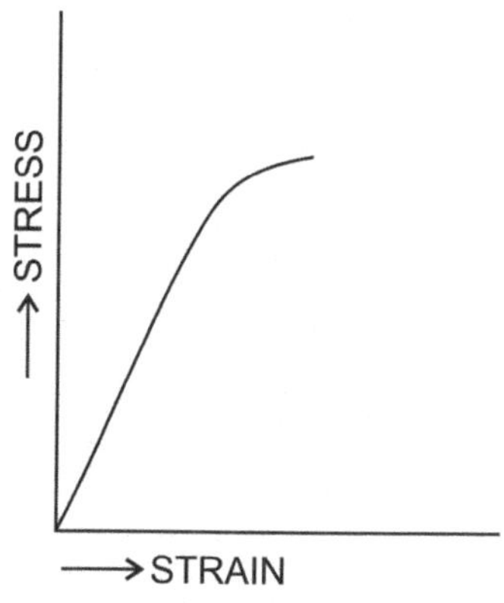

Fig. 1.7 Copper.

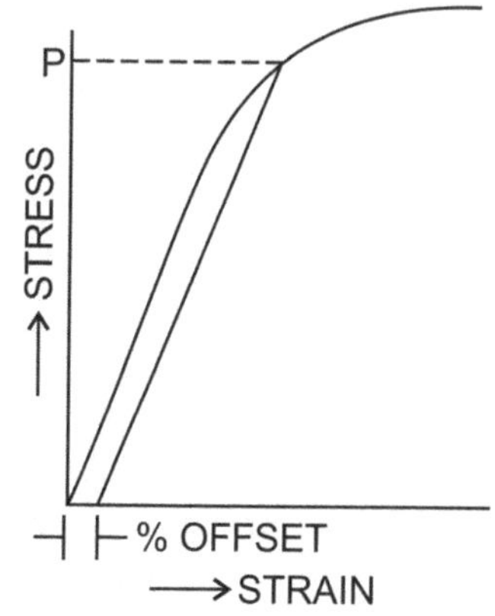

Fig. 1.8 Graphical computation of proof stress.

The most accurate method of determining the proof-stress is only to plot the stress-strain curve very accurately. Usually, a higher load amounting to about 25% of the proof-stress is applied initially. This helps to obtain the elastic portion of the curve at a value higher than zero.

Four-point Method of Calculating the Proof-stress

As it is too laborious and time consuming to plot an elaborate load-extension diagram in routine testing, the four-point method is employed. In this, there is no necessity of plotting any curve. Moreover, it is a fairly accurate method.

This method is aimed at calculating the actual proof-stress, provided the upper and lower proof-stress values are specified. It is the general practice to specify them both. Even if the lower proof-stress is only specified, the upper value may be assumed from our experience with the material. For an easy understanding, the method is explained below with a sketch of a stress-strain diagram (Fig. 1.9).

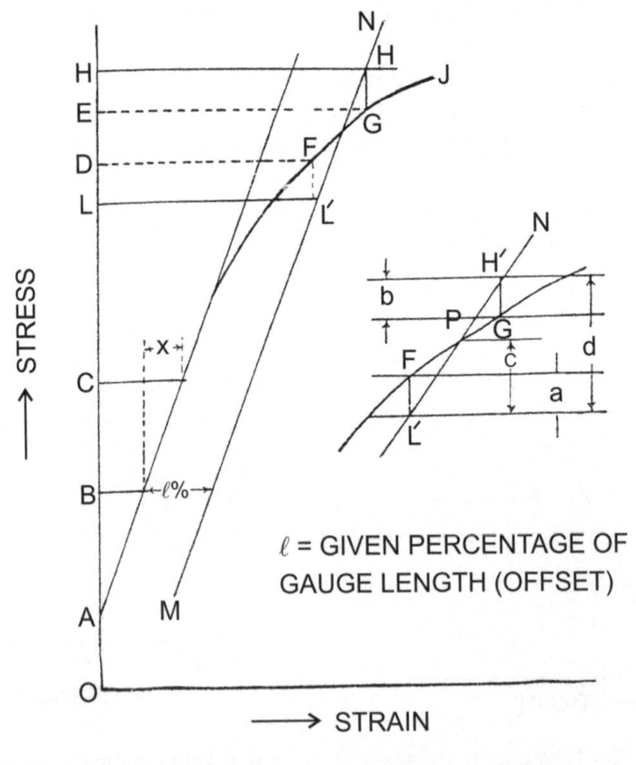

Fig. 1.9 4-point method of determining proof-stress.

The stress-strain curve is shown by *AFGJ*. First, the specimen is subjected to an initial load amounting to *OA* in the diagram. This is about 20% of the* proof-stress of the material. The load is then increased to a higher value, *B*. The extension produced in the specimen is measured. Again, the load is increased to another value, *C*, and the extension measured. In routine testing, this can be directly read by adjusting the extensometer reading to zero at *B*. Let the extension be x.

The extension that will be produced at the value of stress amounting to the lower proof stress can be calculated from the relation:

$$\text{Extension (at L.P.S.)} = \frac{AL}{BC}\,x + l$$

where "l" is the specified proof-stress value of the strain expressed as percentage.

The actual stress representing this value of the strain on the diagram *AFGJ* witlh the point *F*.

Similarly, the point *G* can be located by calculating the extension as per the relation $\frac{AH}{BC}\,x + l.$

Now, if we draw a line MN parallel to the elastic portion of the curve at a distance of l from it, we know that the actual proof-stress value can be found out as the point of intersection of the line *MN* with the curve *AFGJ*. Thus, it can be seen that the actual proof-stress of the material lies some where between the two points *F* and *G* on the curve, viz., above *F* and below *G*.

The actual proof stress can be calculated from the relation:

$$c = \frac{ad}{(a+b)}$$

where d = proof-stress range,

a = the amount by which the stress at *F* is above the lower proof-stress value,

b = the amount by which the stress at *G* is below the upper proof-stress value, and

c = the amount by which the stress at the actual proof is above the lower proof-stress value.

* From the assumed value.

Thus, the actual proof-stress of the material will be the lower proof-stress value plus *c*.

For understanding of the above calculation, the proof-stress region of diagram is separately drawn in Fig. 1.9 as an insert.

In actual experiment, the proof-stress can be simply found out as follows :

(i) After properly fixing the specimen in the machine, a load *A* amounting to 20-25% of the proof-stress of the material is applied.

(ii) The load is increased to about 30-35% of the proof-stress value *B*. The extensometer is adjusted to zero.

(iii) Again, the load is increased to about 40-45% of the proof-stress value *C*. The extensometer is read as *x*.

(iv) The extension at the lower proof-stress value is calculated using the relation $\frac{AL}{BC}x + l$ and the load at which that extension is obtained in the specimen is found out. The value of the load is *F*.

(v) Similarly, the value of the load *G* which is necessary to give an extension equal to $\frac{AH}{BC}x + l$ is found out.

Then, we get

$a = F -$ Lower proof-stress,

$b =$ Upper proof-stress $- G$, and

$d =$ proof-stress range.

(vi) The value of *c* is calculated from the relation $c = \frac{ad}{(a+b)}$ and it is added to the lower proof-stress value. The actual proof-stress is obtained.

THE UNIVERSAL TESTING MACHINE

A photograph of Avery-Denison Universal Testing Machine (model 7113 DCJ) is shown in Fig. 1.10. The top crosshead can be adjusted to three positions for extended tension tests (the left hand side portion of the machine).

The indicator control console contains the pumping unit and the valve gear, the dial indicator and electrical controls. It is connected to the straining unit (right portion of the machine) by flexible pipes and cables. There are two main handwheel controls, one for applying and the other for releasing the load. The loading valve is so designed that at any setting, the rate of oil flow is uniform. When the valve is closed, the load remains stationary long enough for extensometer readings to be taken. The unloading valve is similar and in combination, the two will give all the control needed for applying incremental loads, for applying the loads

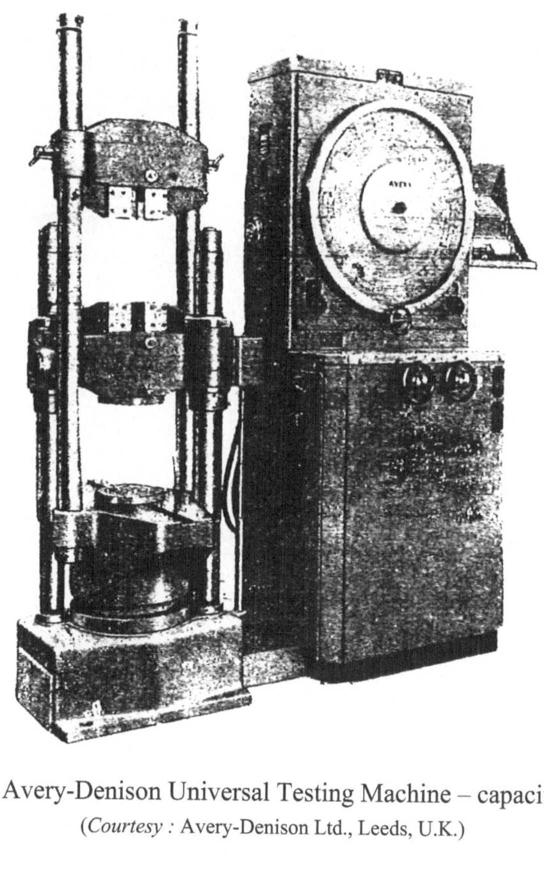

Fig. 1.10 Avery-Denison Universal Testing Machine – capacity 40 K.N.

(*Courtesy :* Avery-Denison Ltd., Leeds, U.K.)

quickly, for holding the loads steady and for removing the loads. An autographic recorder (Fig. 1.11) situated at the extreme left of the machine can be used to plot the stress-strain curve during the test itself. A standard tensile specimen to be used on the machine is shown in Fig. 1.12, while three types specimen gripping devices are shown in Figs. 1.13, 1.14 and 1.15.

Fig. 1.11 Autographic Recorder on Model 7108 Universal Testing Machine. (*Courtesy :* Avery-Denison Lttd., Leeds, U.K.)

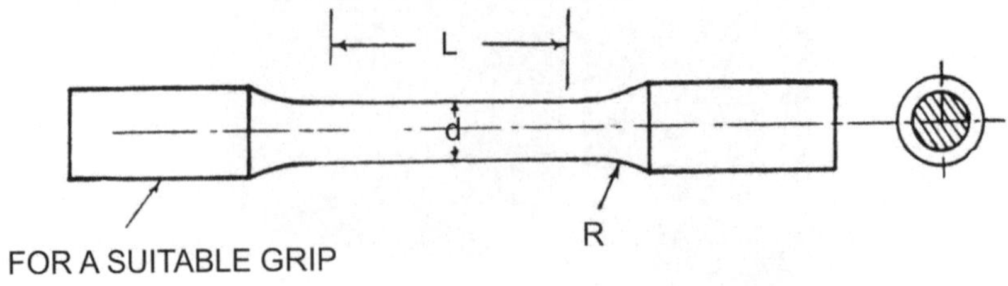

$$L = 5.65\sqrt{d}\ ;\ R = d.$$

Fig. 1.12 Tensile specimen. (All dimensions in mm.)

Fig. 1.13 Specimen gripped using threaded grips in the Universal Testing Machine.
(*Courtesy :* Avery-Denison Ltd., Leeds, U.K.)

Fig. 1.14 Photograph showing a plate specimen gripped in the Universal Testing Machine using pin type of grips.
(*Courtesy :* Avery-Dension Ltd., Leeds, U.K.)

Fig. 1.15 Specimen gripped by wedge grips in a Universal Testing Machine.

(*Courtesy* : Avery-Dension Ltd., Leads, U.K.)

SHEAR AND TORSION TESTS

As briefly explained in the beginning, the shear stress acts parallel to a plane unlike the tensile and compressive stress which acts normal to the plane. The resultants of forces which are parallel, but opposed in direction, act through the sections that are spaced at infinitesimal distances apart. This is an ideal condition of pure shear and the shearing stresses are uniform. A close approximation to this condition is attained in the case of a riveted joint.

If, on the other hand, there is a finite distance separating the two equal and opposite parallel forces, there occurs a tendency to bending in addition to pure shear. A practical example of this condition is that of a beam supported at both its ends (Fig. 1.16). The shearing stress varies from its maximum value at its neutral axis, to zero at the top and bottom layers.

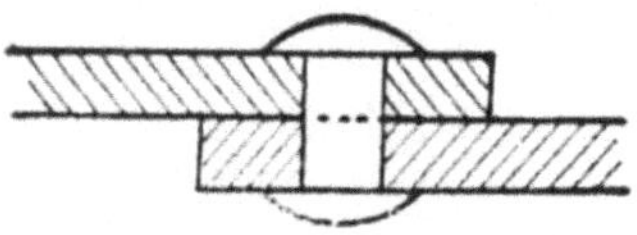

A revised joint illustrating shear.

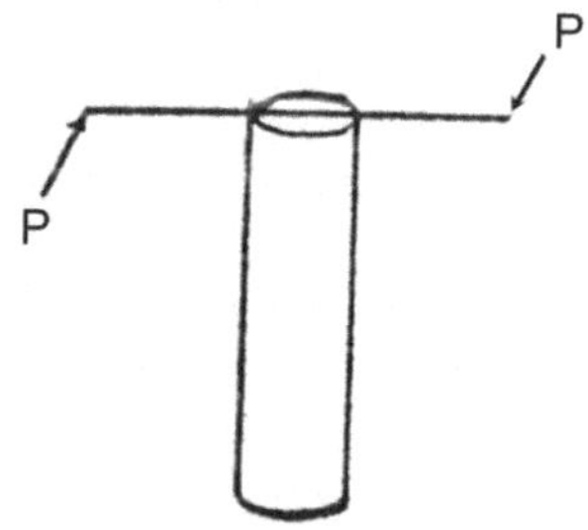

Torsion of a cylinder.

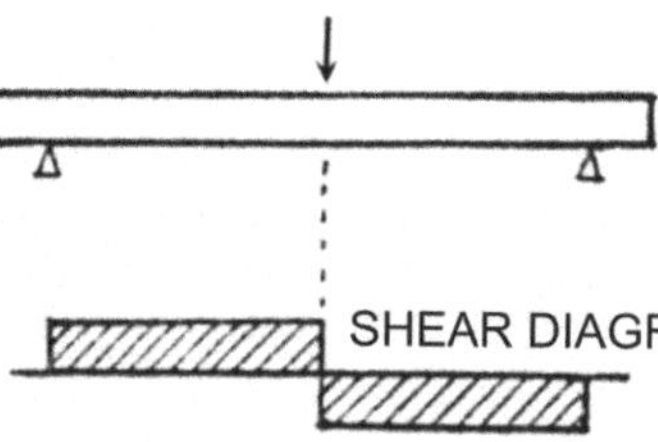

Fig. 1.16 Beam supported at both ends, showing tendency to bend in addition to pure shear.

When the applied forces are parallel and opposite but do not lie in the same plane as the longitudinal axis of the body, a couple is set up. This couple produces a twisting action about the longitudinal axis of the body and this condition is known as torsion.

Torsion Test

The torsion test is not standardized yet. However, the A.S.T.M. has published a tentative recommended practice for cast iron (A 260–42 T). But none of the material specifications require the torsion test. Yet, the value of the torsion test cannot be ignored for it gives us a clear insight in the behaviour of metal parts like axles, shafts and twist drills, as the test approaches the actual service conditions. Torsion tests at elevated temperatures are supposed to give the forgeable nature of the metal. Further, the test enables us to determine the modulus of elasticity in shear, or rigidity modulus and the shearing yield strength.

The test is performed on either a cylindrical bar, tube or pipe. Suitable fixtures are used to prevent the collapse of the ends when testing the tubing or piping. A tubular specimen is preferred as the stress distribution will be uniform throughout the material cross-section. In the case of a cylindrical specimen, the stress will vary from zero at the axis to a maximum at its surface. Again, a very thin wall in the case of a tube is equally undesirable as the specimen may fail by collapse rather than torsion.

The length of the specimen is recommended to be 10 times the diameter (Fig. 1.17). At the same time, the diameter/wall-thickness ratio (in the case of tubes and pipes) should not exceed 10. The testing machine, essentially, consists of a twisting head. The specimen is fixed

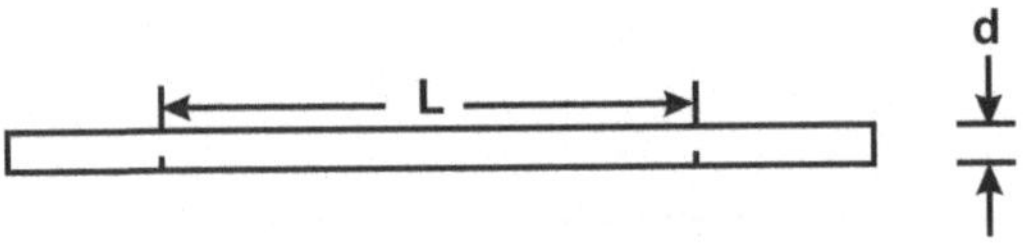

Nominal Dia d in mm		Gauge Length L in mm	
0-4	1.8	200 d	
1-0	5.0	100 d	
5-0	12.5	50	30 d

Fig. 1.17 Standard torsion test specimen.

rigidly at one end and other end is twisted. The torque or the twisting moment applied is measured on a circular scale. The machine has proper centering devices to enable the operator to coincide the axes of the specimen and that of rotation. The deformation in the specimen is noted by two pointers rigidly fixed to the specimen at the two ends of the gauge length (see Fig. 1.18).

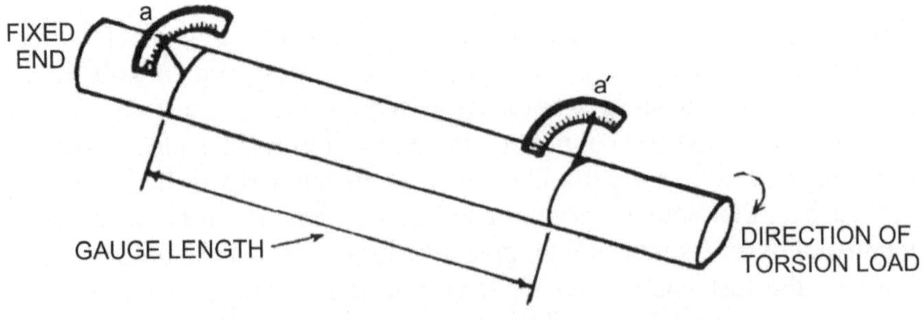

a – Pointer reading at the fixed end of the gauge length.

a′ – Pointer reading at the torsion end of the gauge length.

Fig. 1.18 Principle illustrating the Torsion Test.

The shearing stress can be calculated by the following formula:

$$S = \frac{16\,p_p}{\pi\,d^3} \text{ for cylindrical bars}$$

and

$$S = \frac{16 p_p d}{\pi\left(d_1^4 - d_2^4\right)} \text{ for tubes.}$$

where S = shear stress in Kgf-mm^2,

p_p = torque at lever p in Kgf/mm,

d_1 = external diameter in mm,

d_2 = internal diameter in mm,

d = diameter in mm.

A torsion testing machine is illustrated in Fig. 1.19. This machine has a pendulum weight connected to the pointer of the dial reading the torque. The torque is transmitted through the specimen, on to the pendulum. As a result, the pendulum, which was vertical at the start of the test, is raised to the horizontal position, thus balancing the applied torque. The drive is by either handwheel or electric motor, through a reduction gear.

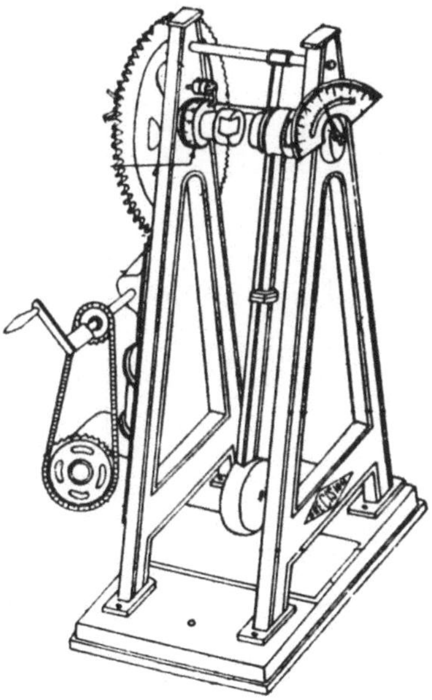

Fig. 1.19 Torsion Testing Machine.
(*Coutesy :* Precision Scientific Equipment Corpn., New Delhi.)

The torsional deformation is measured as angular twist. The net angular deformation is obtained by subtracting the angle noted at the fixed end of the gauge length from the angle noted at the twisted end. This value is divided by the gauge length to obtain the angular twist per mm of gauge length. The shearing strain (in mm per mm) can be computed by multiplying the angular twist per mm of gauge length by half the diameter of the specimen. The modulus of elasticity in shear (rigidity modulus) can be calculated from the equation :

$$D_S = \frac{SL}{r\theta}$$

Where S = maximum shearing strength in kgf/mm^2.

L = gauge length in mm,

r = distance from the axis of the specimen to the outermost fibre, i.e., half the diameter of the specimen in mm, and

θ = angle of twist in radians over a length of L.

Appearance of fracture. It is explained above that torsion is a particular case of shear. The failure occurs at right angles to the axis of the specimen in the case of ductile materials. In the case of brittle materials, however, the axis on which the final twisting took place before the failure has occurred will be clearly visible. The fracture region appears in the shape of a cam (almost similar to a helix), as has been shown in Fig. 1.20.

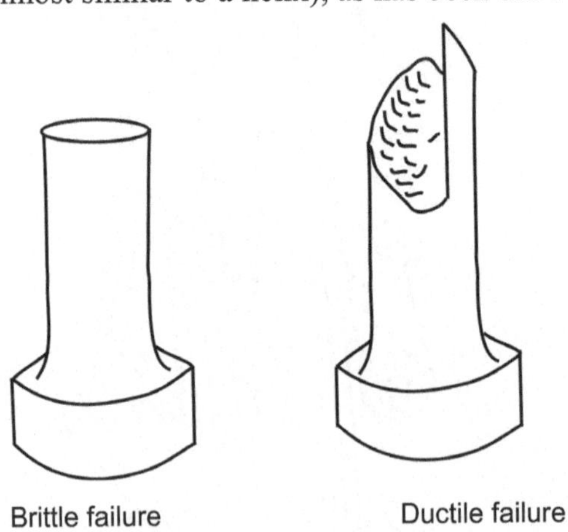

Fig. 1.20 Torsion failures.

The practical application of the torsion test is the near reproduction of the service condition during the test. Power transmission shafts and twist drills are best tested for torsion to determine the rigidity modulus in shear. Laboratory investigations to study the effect of case hardening on steels also require torsion tests. Helical springs tested in compression actually constitute the torsion test. It will be good that the test is standardized and prescribed in greater detail.

Shear Tests

The shear test for metals, though it appears really practical, is neither prescribed nor standardized. Even for rivets, only the tensile strength and cold-bend tests are prescribed, but not shear. The reason is obvious, for a perfect condition of shearing cannot be brought into play during the test, without the overlapping effects of bending and cutting. However, single shear and double shear tests are prescribed with the help of certain fixtures (Fig. 1.21).

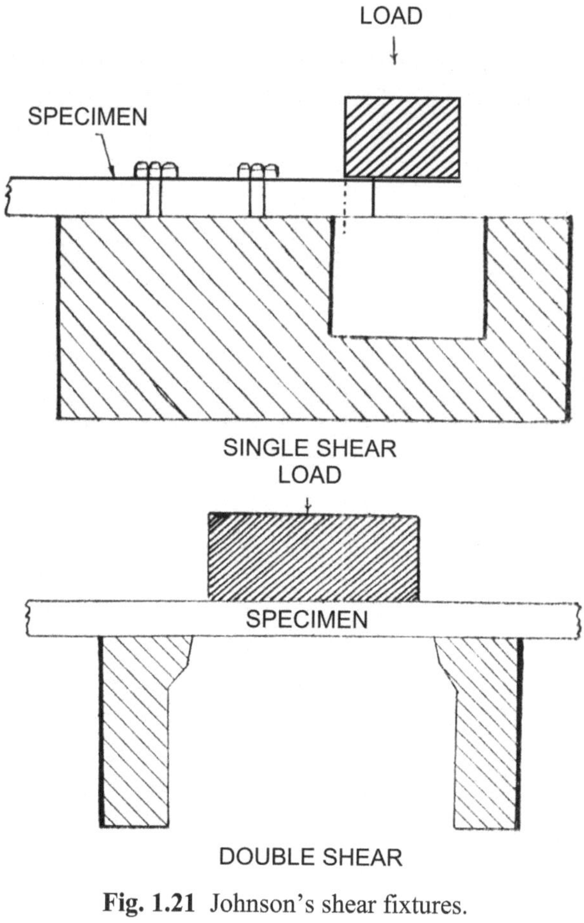

Fig. 1.21 Johnson's shear fixtures.
(*Courtesy :* A.S.M. Metals Handbook, 1948.)

The specimen selected is a 25 × 50 mm section (rectangular section) or a rod of 1″ diameter. It is rigidly clamped to a base. Its ends project out over to a hole. The shearing tool *T* is loaded to rupture the specimen along the shearing plane SS'. The shearing load is measured.

If the specimen extends to the other side of the hole also, and a suitable punch is used to shear, the same test as described above gets modified into that of double shear.

In both the tests, the maximum shear load is noted.

References

IS: 3711-1966, Selection and preparation of test pieces for mechanical tests for wrought steels.

IS: 1608-1960, Tensile testing of steel products other than sheet, strip wire and tube.

IS: 2078-1962, Tensile testing of grey cast iron.

IS: 2654-1964, Method for tensile testing of copper and copper alloys.

IS: 1816-1961, Tensile test for light metals and alloys.

IS: 1717-1961, Simple torsion testing of steel wire.

CHAPTER 2

Hardness

Hardness is a property exhibited by all materials. If we hit a small block of copper with a hammer, the block elongates and a dent of the shape of the hammer tip can be seen on the copper block. From this we can say that the hammer is harder than the copper block. Similarly we can take any two materials and find out which of them is harder. This can be done by hitting one with the other, by rubbing one against another, or trying to pierce or dent one with the other. In all the cases, the harder material remains unaffected. Depending upon the way in which the hardness is found out, the method is suitably termed. When we try to break one of the two materials by sudden loading or impact, it is called impact hardness. The word impact hardness is modified in technical usage and termed as rebound hardness. Particularly for metals which are crystalline, a sudden fracture is not possible but the metal surface gives a rebound to the material which strikes it. The amount of rebound is proportional to the hardness of the surface. In fact, the rebound hardness is measured thus.

Scratch or wear hardness is that property of materials by virtue of which they resist wear or abrasion. Similarly, the amount of resistance offered by a metal for indentation or penetration is defined as the indentation hardness of the metal. Thus, we define hardness as the property of a metal by virtue of which it resists scratch, wear, abrasion or indentation.

In the industry, the indentation hardness is most commonly measured. In the metallurgical sense, it has become a practice to understand hardness as the indentation hardness only, unless otherwise specified. Rebound hardness, too, is often employed for routine check-up. Unfortunately, the three types of hardnesses–indentation, scratch and rebound–are not related to each other or inter-convertible. However, the different indentation hardness values measured are to some extent inter-convertible.

INDENTATION HARDNESS

When we pierce a block of wax and a block of wood with the tip of a nail, the nail will penetrate deeper into the wax than into wood. This is due to the higher resistance offered by wood. The factors which effect the depth of penetration of the nail can be seen to be :

(a) the pressure used (load applied) for indentation,

(b) the dimensions of the tip of the nail, and

(c) the time for which the load is applied.

The depth of penetration can be an index of the hardness of the material if the above three factors are standardized. This is how indentation hardness is measured.

Brinell Hardness Test

Brinell hardness testing machine uses a hardened steel ball as the indenter. Balls of different diameter from 2.5 to 10 mm are employed with suitable loads. A schematic diagram of the Brinell hardness testing machine is given in Fig. 2.1 and a photograph of the same is shown in Fig. 2.2.

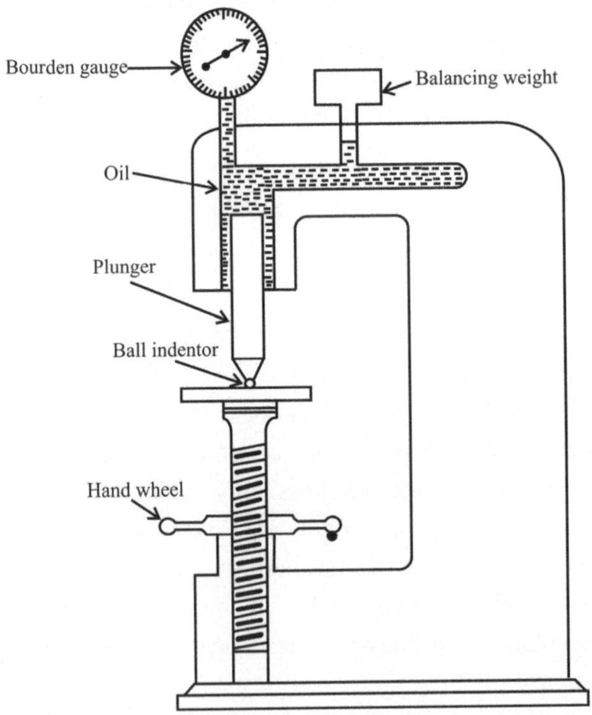

Fig. 2.1 Brinell hardness tester (hydraulic loading). Schematic diagram.

A hydraulic pump applies the load required for the specified time, i.e., 10 secs. The pump is operated by the tester himself in the older type of machines. Nowadays, electrically operated machines are available.

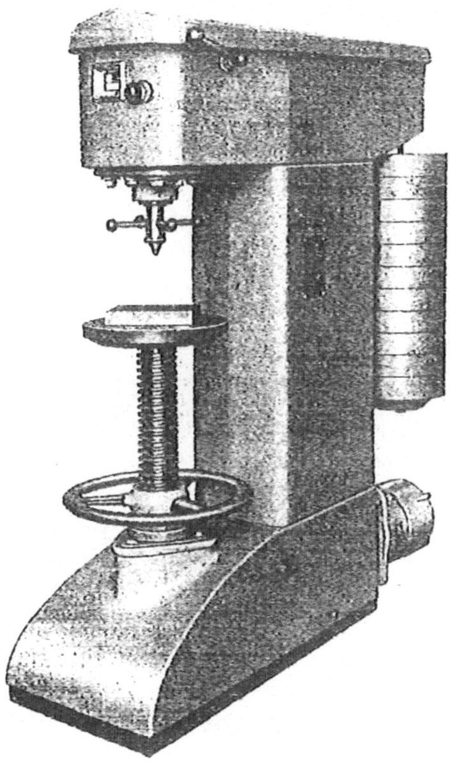

Fig. 2.2 Brinell hardness tester.
(*Courtesy* : Avery-Denison Ltd Leeds, U.K.)

After indenting, i.e., applying the load on the ball, for 10 secs, the load is removed and the indentation made is measured. In the Brinell machine, the surface area of the indentation is calculated and used as an index of the hardness of the metal. The surface area of the indentation is dependent upon the depth of penetration. The load applied (in Kgf) divided by the spherical area of indentation (in square millimeters) is taken as the Brinell Hardness Number.

$$\text{B.H. N.} = \frac{\text{Load } P}{\text{Spherical area of indentation}}.$$

Calculation of the spherical or surface area of indentation. The circle in Fig. 2.3 is the periphery of the indenter, whose diameter is D. The indented portion of the material shown shaded is the replica of a part of the surface of the indenter.

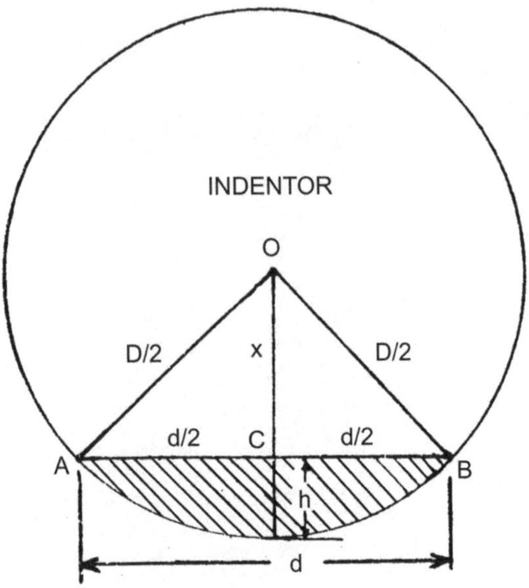

Fig. 2.3 Brinell Indentation

From mensuration principles, the surface area of a segment of a sphere is

$$= \pi Dh \qquad \text{...(1)}$$

where D = diameter of the sphere, and

h = height of the portion under indentation.

From Fig. 2.3, $$h = \left\{\frac{D}{2} - x\right\}. \qquad \text{...(2)}$$

Again, from the right angled triangle OCB,

$$(D/2)^2 = X^2 + (d/2)^2.$$

or $$x = \sqrt{\frac{D^2 - d^2}{4}}$$

$$= \frac{1}{2}\sqrt{\left(D^2 - d^2\right)} \qquad \text{...(3)}$$

Substituting in (2) for 'x' we get,

$$h = \frac{D}{2} - \sqrt{\frac{(D^2 - d^2)}{2}}$$

$$= \frac{D - \sqrt{(D^2 - d^2)}}{2}.$$

Substituting in (1) for 'h', we get the spherical area of indentation

$$A = \frac{\pi D}{2}\left\{D - \sqrt{(D^2 - d^2)}\right\}.$$

The value of 'D' is known and the value of 'd' is measured with the help of a Brinell microscope (Figs. 2.4a and 2.4b). The hardness of the metal can either be calculated from the formula

$$\text{B.H.N.} = \frac{p}{\frac{\pi D}{2}\left(D - \sqrt{(D^2 - d^2)}\right)}$$

or by reading the standard charts which indicate the hardness number directly based upon the load applied and the diameter of the indentation obtained.

Fig. 2.4 a. The Brinell Microscope.
(*Courtesy* : Avery-Denison, Ltd., Leeds, U.K.)

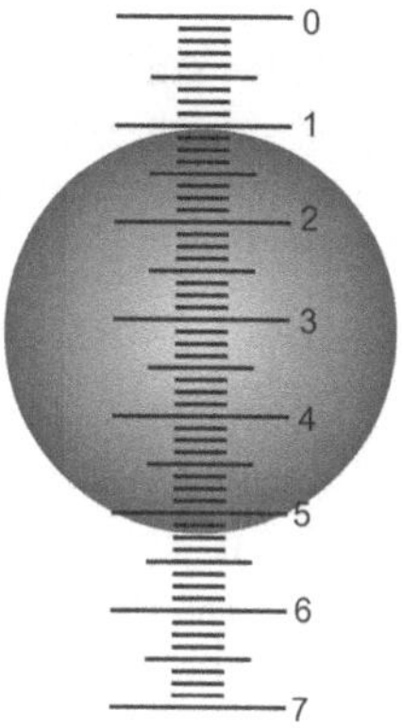

Fig 2.4 b. Brinell Indentation under the Brinell Microscope.
(*Courtesy* : Avery-Denison Ltd., Leeds, U.K.)

Scope and Applicability

The Brinell hardness test makes use of a steel ball as an indenter. The hardness of the steel ball should be sufficiently higher than the material under test, otherwise the indenting ball itself will be deformed. Thus, the hardness of steels in the fully or surface hardened condition cannot be determined by this method. In other words, the Brinell Hardness Numbers are not reliable above 350. Further, depending upon the material under test, right selection of the ball and the load should be made.

From the relation

$$\text{B.H.N.} = \frac{P}{\frac{\pi D}{2}\left(D - \sqrt{\left(D^2 - d^2\right)}\right)}$$

We can see that there is the constant P/D^2,

If this is fixed, the indentations produced or the values of 'd' will be in a particular range. Hence, the ratio P/D^2 is maintained at different values in order to obtain measurable indentalions on the materials.

Usually $P/D^2 = 30$ for steels,

$= 10$ for non-ferrous metals,

$= 2$ for soft metals like lead.

The surface of the material should be even and smooth in order to obtain a clear indentation. Irregularities induce difficulties not only in the proper distribution of the load but also result in impressions with no well-defined contours. Thus, an accurate measurement of the diameter of the indentation becomes impossible.

Very thin specimens should not be tested by the Brinell method. When there is not sufficient material to back up, the region of indentation may become unduly large due to the least resistance offered to the ball. The thickness of the metal under test should be at least equal to the diameter of the indenter.

Similarly, the measurements at the edges and corners of the materials become difficult. Round objects should be filed out at a suitable place so as to obtain a plain surface and tested by placing them on U or V-shaped backing anvil.

The indenting ball should be frequently checked for its spherical shape and accuracy by finding out the hardness of the standard test block supplied along with the machine.

The Brinell hardness number is roughly related to the tensile strength. This empirical relationship is as follows:

$$\text{T.S.} = K \times \text{B.H.N. (tons/inch}^2\text{)}$$

where $K = 0.217$ for alloy steels and

0.22 for plain carbon steels.

For wrought light alloys, the relationship is as follows :

$$\text{T.S.} = 0.25 \times \text{B.H.N.} - 1 \text{ (tons/inch}^2\text{)}$$

The Brinell hardness number (B) is also related to the rebound hardness measured by Shore's scleroscope (S) as per Beeching's formula which is as follows :

$$S = 0.108\,B + 8$$

and $S = 0.1\,B + 15$ when S is more than 55.

[These (Beeching's formulae) are applicable for steel.]

Vickers Hardness Testing Machine

Vickers hardness machine also functions on the same principle as the Brinell hardness testing machine, but employs a square based pyramid made of diamond as the indenter. The size of the indenter is also very small. The included angle between the opposite faces of the pyramid is 136°.

This machine is more versatile than the Brinell hardness tester. Instead of changing the indenter as well as the loads depending upon the nature of the material tested, only the load is changed in the Vickers hardness tester. Varying loads from 1.0 Kgf. onwards up to 120 Kgf. are employed. Owing to the fineness and the small size of the indentation obtained, the specimen needs a glossy surface finish for testing. Generally the loading device is in the form of a pedal. The indenter and an objective to measure the indentation can be swung into the correct positions as required. A schematic diagram of the machine is given in Fig. 2.5.

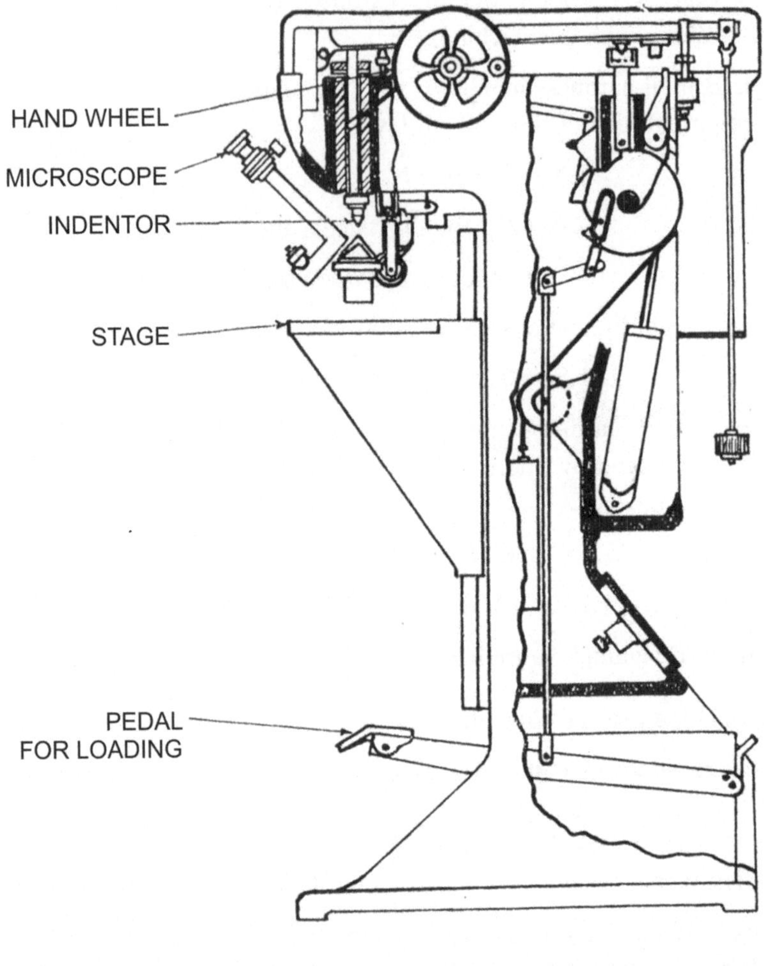

Fig. 2.5 Vickers hardness testing machine. (Schematic diagram.)

After indentation the size of the indentation is accurately measured by swivelling the microscope into position. The impression obtained will have a square sectional area (Fig. 2.6). The diagonal is measured using a crosswire focusing device in the optical equipment. In some types of hardness machines like the Galileo type, the indentation can be focused onto a graduated ground glass screen and measured.

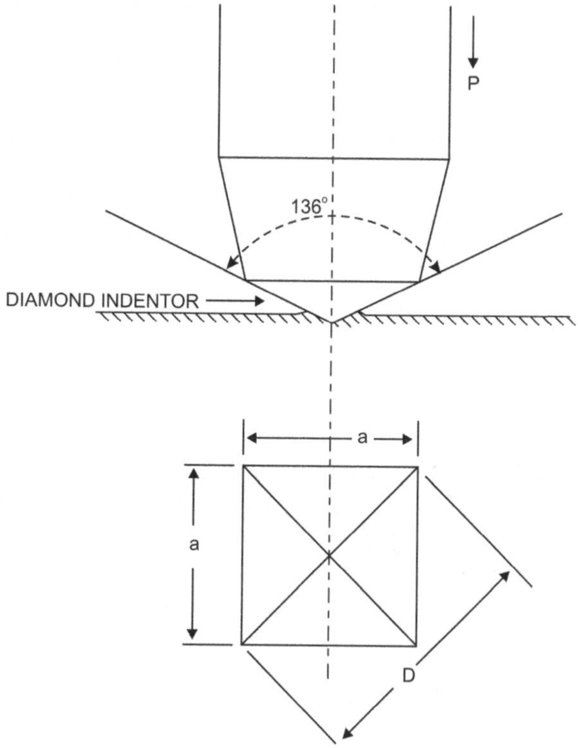

Fig. 2.6 Vickers indentation.

As in the case of Brinell hardness tester, the numerical value of Vickers hardness number is equal to the load applied, divided by the surface area of the pyramidal indentation. Thus the diamond pyramid hardness number

$$\text{D.P.H.} = P/A$$

where P = load applied, and

A = lateral area of pyramidal indentation in sq. mm.

The area of the pyramidal impression is calculated from the mensuration formula.

$$A = 4a\,S/2 \qquad ...(1)$$

where 'S' is the slant height of the impression.

From Fig. 2.6 it can be seen that

$$D^2 = 2a^2 \text{ or } a = D/\sqrt{2}$$

Thus, $A = (4D/\sqrt{2})\,(S/2)$

$$= \sqrt{2}\,DS \quad \text{...(2)}$$

The included angle of the indentation is the same as the included angle of the indenter pyramid. This angle is 136° (Fig. 2.6). The slant height can then be expressed in terms of D as follows:

$$S = \frac{a}{2\ \sin\ 68}$$

$$= \frac{\dfrac{D}{\sqrt{2}}}{2\ \sin\ 68} \quad \text{...(3)}$$

Substituting the value of 'S' from (3) in (2) we get

$$A = \frac{D^2}{2\ \sin\ 68} \quad \text{...(4)}$$

If the value of 'A' as for (4) is substituted in the hardness number equation. Then,

$$\text{D.P.H.} = \frac{P}{D^2/2\ \sin\ 68}$$

$$= \frac{2.\ \sin\ 68.\ P}{D^2}$$

If the value of the constant 2 sin 68 is evaluated,

$$\text{D.P.H.} = 1.854\ P/D^2$$

where P = applied load,

and D = diagonal length of the impression.

Standard charts are supplied along with the Vickers machines, which give the Vickers hardness number depending upon the diagonal length of the impression at each load. Using these charts, the hardness number can be readily found.

The machine is the most accurate and versatile. Very high degree of polishing of the metal surface is required so that the impression can be perfectly focused and viewed and the diagonal accurately measured. Very careful handling of the indenter is recommended, besides periodic checking of the instrument against the standard steel blocks supplied alongwith.

Brinell cum Vickers Hardness Tester

Of late, machines capable of performing either Brinell or Vickers hardness tests are in the market. One such, Avery Denison (Type 6406), is illustrated in Fig. 2.7 a. It is supplied with hardened steel ball indentors of 1 and 2 mm diameter as well as the pyramidal diamond indentor. The indentation load can be varied by changing the weights at the back of the instrument. The specimen surface should be mirror-polished so that the indentation is accurately measured in the projected image on the ground glass screen.

Fig. 2.7 a. Model 6406 Vickers and Brinell hardness tester.
(*Courtesy* : Avery-Denison, Ltd., Leeds, U.K.)

The indenter as well as the image projection objective will come into position while loading and measuring, respectively, by means of the electric make-and-break mechanism inside the instrument. As usual, the machines are supplied with all the accessories like an anvil of larger diameter (to hold larger test pieces), slotted anvil (to support round stock), adjustable work rests (for lengthy objects), and calibrating (standard) Hardness blocks. The indentation is magnified and projected onto a ground glass screen fitted at the top of the machine. A vernier mechanism, provided in the screen, facilitates accurate measurement of the indentation (Fig. 2.7 b).

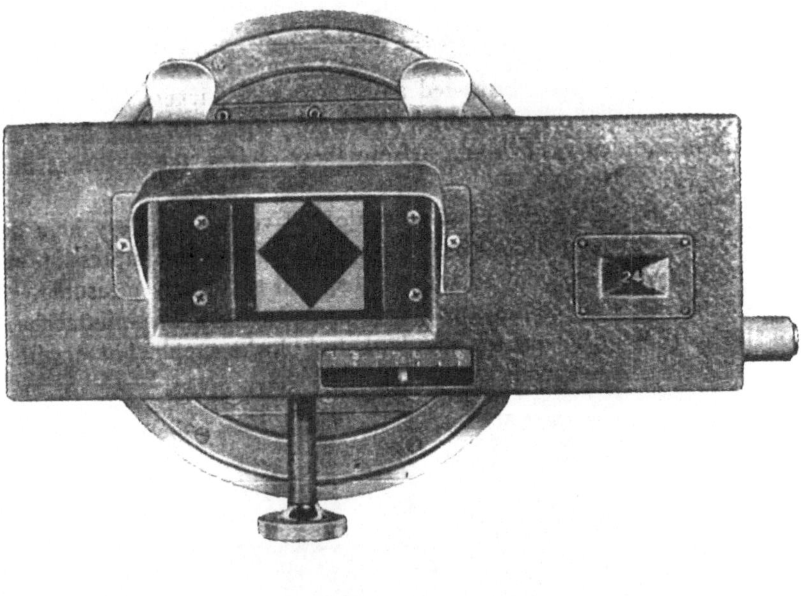

Fig. 2.7 b. Measuring screen showing magnified image of Vickers impression.

The Brinell and Vickers hardness values are similar in their nature from the basic principles. The two values will tally up to the hardness number 400, and thereafter, the Vickers values are found to be more (see Fig. 2.8). However, in this range, i.e., above 400 Vickers hardness values are more reliable, as already mentioned above.

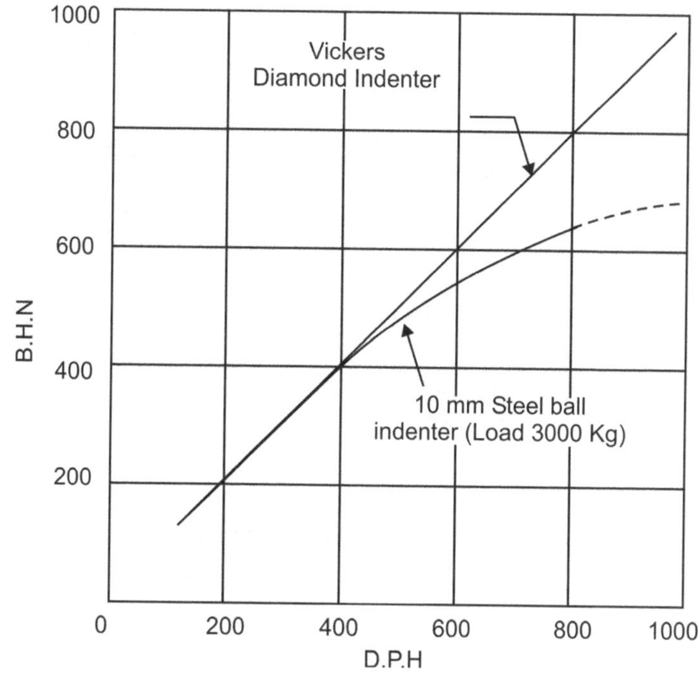

Fig. 2.8 Comparison of Brinell and Vickers hardness values.

Rockwell Hardness Tester

We have seen hitherto that when a metal is indented, it opposes the indentation. The degree of resistance offered by the metal is commensurate with its hardness. If the hardness is more, the resistance to indentation is more, and vice versa. Thus, we will have small indentations with hard metals and larger indentations with soft metals. Further, we have computed the surface area of indentation as an indirect index of the hardness.

Based on the same phenomenon, we can see that the depth of penetration is also inversely proportional to the material hardness. This can be utilized as a means of measuring the hardness. This concept was proposed in 1908 by Ludwig at Vienna.

Rockwell hardness tester is developed with the depth of penetration as the criterion for the hardness of the metal. A schematic diagram, underlining the principle of the working of the instrument, is given in Fig. 2.9.

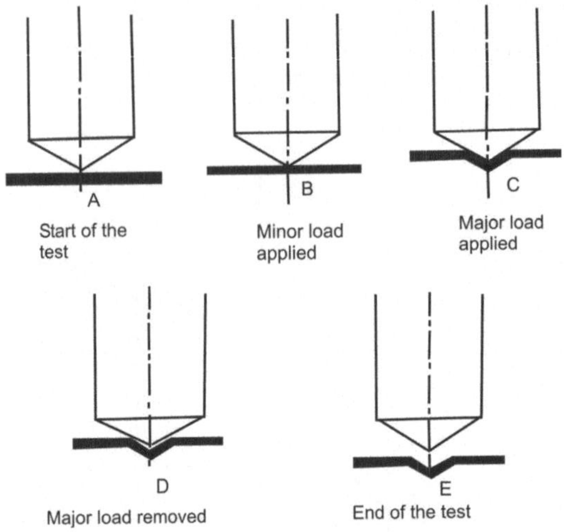

Fig. 2.9 The Rockwell indentation procedure.

Description of the machine. The photograph of a common type of Rockwell hardness tester is shown in Fig. 2.10. The machine consists of

Fig. 2.10 Special Rockwell for large parts.
(*Courtesy* : Wilson Instrument Division of A.C.C.O, Connecticut, U.S.A.)

an anvil which can be changed depending upon the shape of the specimen under test. Different anvils are illustrated in Fig. 2.11. The anvil can be moved up or down by turning the hand wheel which is situated at the bottom of the spindle. A loading lever is situated at the right hand side bottom portion of the machine. The loading may also be applied by simply operating a hand-lever which is just below the handwheel.

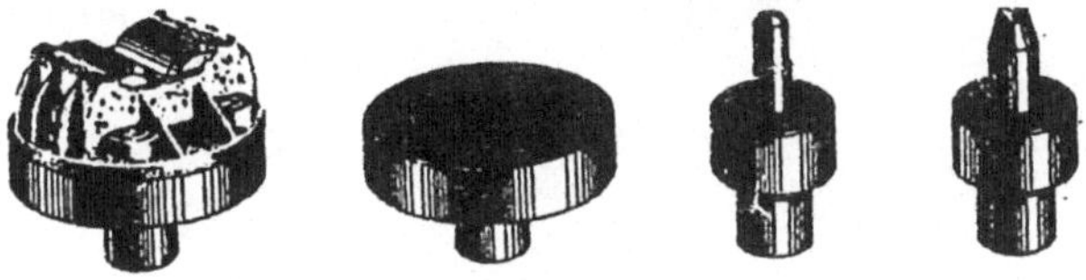

Cylindron Jr. **Plane** **Spot** **Shallow V**

Fig. 2.11 Different types of anvils.

(*Courtesy* : Wilson Instrument Division of A.C.C.O., Connecticut, U.S.A)

The machine is a direct reading type. The dial consists of two pointers. The small pointer which is situated in the top left quarter of the dial proper indicates the application of the minor load of 10 Kgf. The anvil, with the specimen placed on it, is turned slowly so as to lift it upwards. The small pointer starts to move, once the specimen touches the indenter. The anvil is lifted up until the pointer comes to the stipulated position, viz., until it comes up to the black dot. This indicates that the minor load of 10 kgfs is acting upon the indenter.

The dial is then turned to the required position, i.e., the zero on the C-scale. The major load is applied by pressing the hand-lever (Fig. 2.9C).

Under the influence of the major load which is an increment over the already applied minor load, the indenter starts to go down into the specimen during the specified time. This can be seen from the dial. The pointer starts to move during the period of loading. After the pointer stops, the major as well as the minor loads are removed by turning the hand-lever at the right-hand side bottom in an anticlockwise direction. The reading on the relevant scale is the hardness of the metal (Fig. 2.9D).

The action of the indenter under the influence of minor and major loads is schematically illustrated in Fig. 2.12.

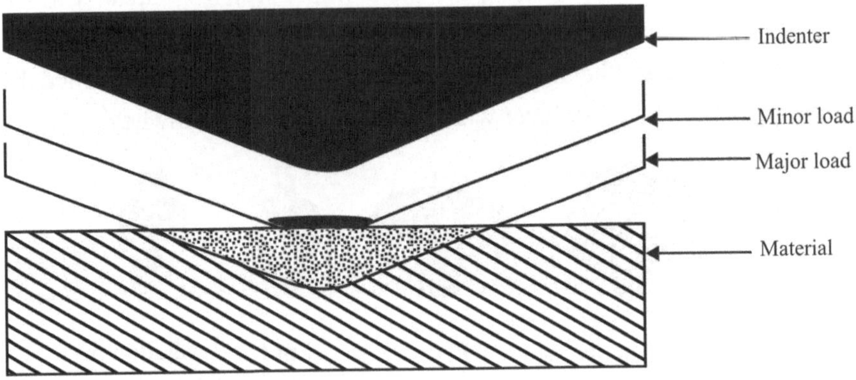

Fig. 2.12 Rockwell indentation.

The indenters and scales. The common indenters employed in the Rockwell hardness tester are a 1/16″ hardened steel ball and a diamond cone. When the former is employed, the major load is maintained at 100 Kgf and the hardness values read on the B-scale. The readings range from zero to 100. The B-scale is suitable to test plain carbon steels, and common non-ferrous metals and alloys like copper, brasses, bronzes, etc. However, for very soft metals like lead, indenter balls having 1/8″ , 1/4″ and 1/2″ diameters are also supplied. When these balls are used, the major loads also will be altered and the hardness value is expressed on a different scale, e.g., instead of R_B, it is expressed as R_E when the 1/8″ diameter, ball indenter is used. It means that hardness value is designated by the E-scale. For testing very hard materials like hardened steels a diamond indenter is employed. It consists of a 120° conical-shaped diamond with a 0.2 mm diam. spherical indenting tip. This is also known by the patented name 'Brale'. With Brale, the major load consists of 150 Kgf. The hardness is read on the C-scale. A detailed account regarding the various hardness scales employed in the Rockwell hardness tester is given in a tabular form.

Rockwell Hardness Scales and Prefix Letters.*

Scale symbol and prefix letter	Indenter	Major load, Kg	Dial Numerals	Typical applications of scales
B*	Group one** 1/16-in. ball	100	Red	Copper alloys, Soft steels, aluminum alloys, Malleable iron.
C**	Diamond cone	150	Black	Steel, hard cast iron, pearlitic malleable iron, deep case-hardened steel
A	Group two Diamond cone	60	Black	Cemented carbides, thin shallow case-hardened steel
D	Diamond cone	100	Black	Thin steel, medium case-hardened steel
E	1/8-in. ball	100	Red	Cast iron, aluminum and magnesium alloys, bearing metals.
F	1/I6-in. ball	60	Red	Annealed copper alloys, thin soft sheet metals
G	1/I6-in. ball	150	Red	Phosphor bronze, beryllium copper, malleable iron.
H	1/8-in. ball	60	Red	Aluminum. Lead, zinc
K	1/8-in. ball	150	Red	Bearing metal and other very soft or thin materials. Use smallest ball and heaviest load that does not give anvil effort.
L	Group three 1/4-in, ball	60	Red	
M	1/4-in, ball	100	Red	
P	1/4-in, ball	150	Red	
R	1/2-in, ball	60	Red	
S	1/2-in, ball	100	Red	
V	1/2-in, ball	150	Red	

* Based on ASTM E 18.

** Commonly used scales and indenters.

Both the most commonly used Rockwell scales B and C consist of 0-100 divisions. But they are staggered in such a fashion that the B 30 is at C 0. It is made so, to avoid the negative hardness values on the B-scale if used to test very soft materials. This also facilitates in establishing that the highest hardness that can be measured with a 1/16″ diam. ball indenter is only B_{100} and for higher hardness the C-scale should be employed.

The actual usable range of the Brale is only from C_{20} onwards up to C_{70}. C_{20} corresponds to B_{97} and below that hardness level, it is not recommended to employ the Brale. In such cases, the shape irregularities at the base portion of the diamond cone begin to play an important role in giving false hardness values. Much care may not be exercised in shaping the bottom regions of the Brale as is exercised on the circular tip. On the other hand, if the C-scale is used for testing materials which are harder than C_{70}, the diamond cone may be damaged due to the high value of the load (150 Kgf) and the reaction of the extremely hard surface. In such instances the A-scale which employs a smaller load of 60 Kgf is better suited.

The useful range of the C-scale is practically from 40-70. This corresponds to the Vickers hardness number range of 410 to 1000. On the other hand when the A-scale is employed, the useful range is between 63-90, roughly corresponding to the Vickers hardness number range of 240 to 1400.

Advantages of the process

Listed below are the advantages of the process :

1. The process is relatively fast.
2. The hardness is directly read on the dial.
3. There is greater latitude for soft to hard materials.
4. A small indentation is left on the object.
5. Use of the initial minor load avoids the errors arising out of the uneven surface of the metal, and machine and table errors.

Disadvantages

The process has its drawbacks which are as follows :

1. The scale range is contracted.
2. The use of more than one scale to express the hardness.

3. Difficulty to readily convert the Rockwell hardness values either into Brinell or Vickers.
4. Errors are likely to be induced when the work is large and overhangs.

Rockwell Superficial Hardness Tester

This machine, shown in Fig. 2.13, is similar to the ordinary Rockwell machine. The indenters used are also the same, viz., the 1/16″ hardened steel ball and the conical based spherical diamond indenter. For softer materials the ball is used and for harder materials the Brale is used. When used in the superficial tester the Brale is designated as 'N Brale'.

Fig. 2.13 Rockwell superficial hardness tester.

(*Courtesy :* Wilson Instrument Division of A.C.C.O; Connecticut, USA)

The hardness reading dial too is similar to that of the ordinary machine. It consists of 100 equal divisions. Each division represents 0.001 mm vertical motion of the penetrator. The dial is very sensitive to the motion of the indenter as the impressions obtained are very small.

Rockwell Superficial-Hardness Scales*

	Scale symbols				
Major load, kg	***N* scale, Diamond cone**	***T* scale, 1/16-in. ball**	***W* scale, 1/8-in. ball**	***X* scale, 1/4-in. ball**	***Y* scale, 1/2-in. ball**
15	15 *N*	15 *T*	15 *W*	15 *X*	15 *Y*
30	30 *N*	30 *T*	30 *W*	30 *X*	30 *Y*
45	45 *N*	45 *T*	45 *W*	45 *X*	45 *Y*

*Based on ASTM E 18.

A minor load of 3 Kgf is used and the major load may be 15, 30 or 45 Kgf as against the 10 Kgf and 60, 100 and 150 Kgf, respectively, in the ordinary machine. Because of this, superficial hardness scales are devised. These are defined by the indenter used. Thus when the N Brale is used the hardness is read as a N-number. Besides, the major load employed is also mentioned. Thus, if the N Brale is used with a 30 Kgf. major load the scale is prefixed as 30N and so on.

Standardization

The machine is frequently checked by finding the hardness of the standard plates supplied along with it.

Application

The machine is particularly suitable when the hardness of the metallic region to be determined is shallow such as case-hardened cases, electroplatings, etc.

Monotron Hardness Tester

In this machine a standard 0.75 mm hemispherical diamond indenter is employed. It is forced into the metal upto a standard depth of 0.05 mm. The load required to force the penetrator is dependent upon the hardness of the material.

The machine consists of two circular scales. The top dial indicates the load. The hardness number is read directly on the top scale with the indenter still in the specimen. The scale is graduated in Kgf. and Brinnell units. For testing softer materiats, large indenters made of tungsten carbide, in diameters of 1.53 or 2.5 mm. are available. When these are employed, the hardnesses are referred to as of the M_3 and M_4 scales respectively.

The bottom dial is for reading the depth of penetration. There are two scales on the dial, each being divided into 100 equal parts. One scale, which is marked in the clockwise fashion, is for routine testing, i.e., the constant depth indenting. The other scale, marked in the anticlockwise direction, is for performing specific tests, such as measuring the depth of indentation or flow of the metal at a constant load of indentation.

The machine can also be used to test the hardnesses of thin layers like the carburised or nitrided cases and thin strip. In such cases where a deeper impression is undesirable, the penetration may be done up to a fraction of the specified 0.0018″, and the noted hardness value multiplied by the numerical in the denominator of the fraction. If an indentation of 0.0009″ depth is made and the load noted, the hardness is evaluated as the load multiplied by 2.

Instead of the Knoop indenter, an ordinary Vickers type pyramid may also be employed for indentation. In such a case the hardness calculation is done as in the Vickers hardness machine.

Precautions

The instrument is very sensitive and every care should be taken to see that the measurment of the diagonal is very accurately done. Since only the projected area of the indentation is taken into account, the measurement should be made exactly at right angles to the surface under test. To achieve this, the bottom portion of the specimen too should be very carefully prepared so that it rests properly on the stage of the instrument.

Mayer's Hardness

It is more rational to consider the projected area of indentation than the actual surface area in the calculation of the hardness. This idea was put forth by Meyer in 1908. According to him, the following empirical relationship exists between the applied load and the size of the indentation,

$$P = Kd^{n1}$$

where $P =$ load applied in Kgs,

$K =$ a material constant based upon the resistance of the metal,

$n^1 =$ a material constant based upon the strain hardening of the metal,

$d =$ diameter of the indentation in mm.

If log p is plotted against log d, a straight line graph is obtained. The slope of the straight line is n^1 and K is the value of P when $d = 1$. The value of n^1 for fully annealed metals is found to be about 2.5 and for fully strain-hardened metals, about 2.0.

It can be proved that the value of P/d^2 is a constant in this case also, i.e., when the projected area of indentation is considered instead of the surface area for hardness purposes. The manufacture of this type of hardness testers is not in vogue now.

Tukon Hardness Tester

This is a special purpose machine, used for determining the hardness of extremely thin materials. Nitrided or cyanided cases and decarburised surfaces of steel and the microconstituents in an alloy can be tested by this instrument. Commonly, the depth of penetration of the indentation does not exceed one micron in this test.

The instrument shown in Fig. 2.14 is a complicated yet precision built apparatus. Except the measurement of the indentation, all the operations

Fig. 2.14 Tukon hardness tester.
(*Courtesy* : Wilson Instrument Division of A.C.C.O., Connecticut, U.S.A.)

are performed automatically by electronic control devices. To test the hardness of a microconstituent in an alloy, the polished and etched specimen is placed on the stage of the instrument. The phase under test whose hardness is being determined, is accurately focused and the specimen is lifted up by moving the stage upwards just below the indenter. Then the instrument is put on by pressing the button. The automatic mechanism functions and performs the following operations:

(i) further movement of the specimen to touch the indenter,

(ii) stopping the upward movement and applying the load for the required time,

(iii) upward movement of the indenter after loading, and

(iv) lowering of the stage and thereby the specimen to the initial position.

The Knoop indenter, which is generally used, is a typically shaped diamond pyramid. The pyramid is cut to an included longitudinal angle of $172^{o}\ 30'$ and transverse angle of $130^{o}\ 0'$, within a 1% tolerance. This facilitates in obtaining an impression which is rhombic in shape with the longer diagonal approximately seven times the shorter diagonal. The indenter and the impressions obtained are shown in Fig. 2.15.

Hardness Test	Brinell	Rock well		Vickers	Knoop	Monotron
Material of the indenter	Steel	Diamond	Steel	Diamond	Diamond	Diamond
Indenter shape	Sphere	Cone	Sphere	Square pyramid	Rhombo pyramid	Hemi sphere
Dimensions of the indenter	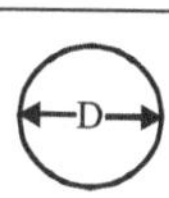 D = 10	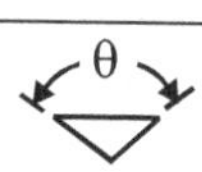 $\theta = 120^{o}$	 D = 1/16″ 1/18″ 1/4″ or 1/16″	 $\theta = 136^{o}$	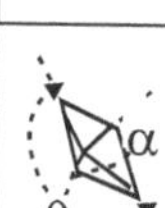 $\alpha = 130^{o}$ $\theta = 172^{\circ}30'$ $172^{o}\ 30'$	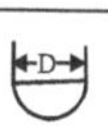 D = 0.75

Fig. 2.15 a. Specifications for hardness indenters.

(From A.S.M. Metals Handbook, 1948.)

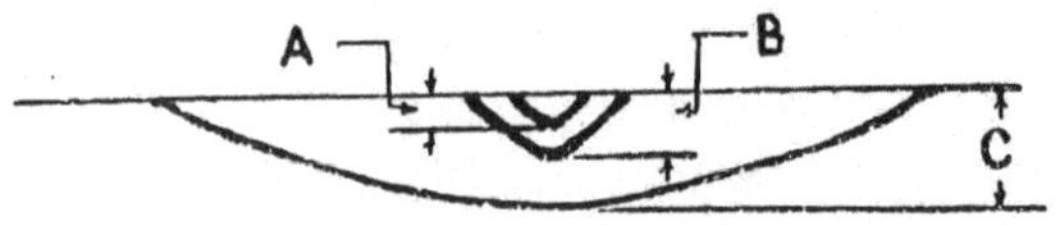

A. Superficial rockwell N diamond cone 30 Kgf. Load 0.0018″

B. Rockwell C scale brale 150 Kgf. Load 0.0052″

C. Brinell 10mm. Ball 3000 Kgf. Load 0.01″

Fig. 2.15 b. Comparative impressions in steel (R_e 39) using Brinell, Rockwell and Rockwell Superficial Testers. (Magnified approximately 30 times.) (From Lysaught, V.E., Indentation Hardness Testing, Reinhold, N.Y., 1949.)

The longitudinal diagonal is the one measured in the impression. The numerical value of the Knoop hardness number is given by the formula:

$$\text{Knoop Hardness Number} = P/A_p$$

where P = applied load in Kgf.

A_p = projected area of the indentation.

The projected area of the impression in the case of an indenter with perfect included angles is equal to

$$wl/2$$

where w = length of the shorter diagonal,

and l = length of the larger diagonal.

From the consideration of the included angles,

$$\frac{\tan 65^\circ}{w} = \frac{\tan\ 86^\circ\ 15'}{l}$$

Substituting the numerical values,

$$l = 7.114.w.$$

or $$w = 0.14056\ l$$

Thus $$A_p = \frac{0.14056\, l^2}{2}$$

$$= 0.07028\, l^2$$

The value, 0.07028, is known as the Knoop indenter constant C_P. This denotes that if the indenter in use is having slight dimensional deviations, the actual value for C_P should be used rather than the theoretical value of 0.07028. The value of C_P is supplied by the manufacturers.

Thus $$\text{KHN} = \frac{P}{C_P L^2}$$

where KHN = Knoop hardness number,

P = load in Kgf,

L = the longitudinal diagonal of the impression, and

C_P = Constant supplied by the manufacturers of the machine.

Instead of using the Knoop indentor, an ordinary Vickers type pyramid indentor may also be employed in the machine. In such a case the hardness calculation should be performed as in the case of the Vickers hardness testing.

Precautions

The instrument is very sensitive and great care should be taken to see that the measurement of the diagonal is made accurately. Since only the projected area of the indentation is taken into account, the measurement should be exactly at right angles to the surface under test. To achieve this, the bottom portion of the specimen should also be very carefully prepared, so that it rests properly on the stage of the instrument.

Rockwell Microficial Hardness Tester

The principle of the Rockwell microficial hardness tester is based on the well-known Rockwell test. The loads applied are much lighter and the depth measurement system is more sensitive in view of the shallow indentations (as shallow as $2\mu m$). This shallow indentation is the result of the microficial indentor which is basically a Vickers indenter with an accurately ground flat tip. This flat tip limits the total depth of penetration without a loss in sensitivity. The principle of working is illustrated in Fig. 2.16.

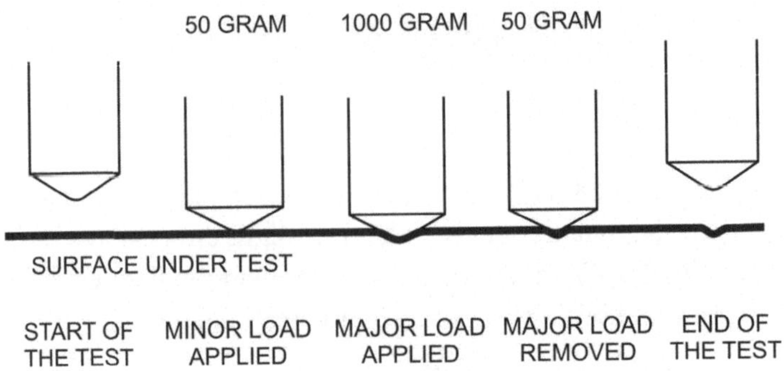

Fig. 2.16 Rockwell microficial hardness test.

The microficial tester is shown in Fig. 2.17. The equipment is the latest addition to the microhardness instruments and is of a direct reading type. Further, the whole cycle of testing takes just 15 seconds.

Fig. 2.17 Rockwell microficial hardness tester.
(*Courtesy* : Wilson Instrument Division of A.C.C.O., Connecticut, U.S.A)

Accuracy of the Rockwell Hardness Tester

A detailed survey was undertaken by a sub-committee of the British Standards Institution in the early sixties. It was reported that every three out of four Rockwell hardness testers working in the industry were giving non-repeatable values of hardness. Later on, the National Physical Laboratory (U.K.) had undertaken detailed investigation and came to the conclusion that the reasons for the lack of repeatability of hardness values of the industrial machines were poor maintenance of the machines and improper testing methods like not supporting the work under test properly. It was concluded that the principle of Rockwell testing was inherently a very accurate method.

It is prescribed by the British Standards Institution that in periodic checking with standard blocks, the machines should give hardness values in the range of ± 2 readings for the B-scale and ± 1.5 readings for the C-scale.

The Indian Standard Specification I.S. 3804 : 1966 stipulates that the machine should give readings with standard blocks within 3% of the actual hardness (of the block) for the C-scale and 6% for the B-scale.

Scratch hardness

This is by far the oldest hardness testing method. Though seldom used except for mineralogical purposes, the testing is very simple according to a hardness scale known after Mohs. It consists of ten minerals arranged in increasing order of hardness. The principle behind this method of testing is that when two materials are scratched against each other, the harder of the two scratches the other. The material under test is scratched with the mineral number 1. If the mineral fails to scratch the material, then the next harder mineral, i.e., mineral number 2, is used, and so on until the material is scratched by a mineral. Suppose the mineral which scratches is the one numbered 5. Then the hardness of the material is expressed as lower than 5 and higher than 4. The Mohs scale of hardness is as follows:

1. Talc
2. Gypsum
3. Calcite
4. Fluorite
5. Apatite
6. Orthoclase (or feldspar)
7. Quartz
8. Topaz
9. Corundum
10. Diamond.

Small pieces, of about two cm. size, are arranged in numbered chambers in a wooden box resembling the weight box of a conventional balance. This forms an important appliance for testing in geological laboratories, besides serving as a portable unit for geological expeditions.

File Test

This is a very useful method of testing the abrasion hardness of metals. A workshop file is all that is used. The material under test is just filed and it is noted whether the file cuts it or not.

On the Mohs scale, the hardness of a file can be put at 6.

An important point to be noted here is that, undue pressures of filing should be avoided, or else the material as well as the file may be cut off.

Penknife Test

An ordinary penknife which is commonly placed in the key purse is used to scratch the material in this test. The hardness of the penknife can be put in the range of 4 to 6 on the Mohs scale.

The Microcharacter Hardness Test

The instrument appears like an ordinary metallurgical bench microscope. In fact, it is so in that the whole thing is an attachment made to a metallurgical microscope.

To test the scratch hardness of a microconstituent, the polished and etched specimen is placed on the stage of the instrument. The exact phase or microconstituent to be scratched is focused and the microcharacter attachment is fixed to the objective. The whole attachment is simple and consists of a well-balanced lever to the bottom of which a sharp diamond-cutting tool is fixed. The lever arm is properly weighted and lowered on to the surface under test. When the micrometer arrangement of the microscope stage is moved, the specimen rigidly fixed on to it also moves against the diamond pointer. Consequently a scratch is made on its surface. The width of the scratch (made on the surface) is governed by the hardness of the material. The width of the scratch differs from region to region, it being wider on the regions which are soft and narrow on the regions which are hard.

Watch oil is used to coat the diamond before cutting so that the cutting action will be smooth. The specimen is removed, cleaned with an organic solvent (cleaning reagent like xylol or carbon tetrachloride) and the width of the scratch on the required phase is measured.

The microhardness 'K' of the phase is obtained by the following empirical formula:

$$K = \frac{10,000}{\lambda^2}$$

where λ is the width of the scratch in microns.

The figure 10,000 is just a conventional one, to avoid fractional values of 'K'.

Generally the load applied to the diamond is 3 gmf. The diamond cutter is attached to a spring sort of a device so that it will scratch the surface effectively, especially when relatively harder and softer regions are present side by side.

REBOUND HARDNFSS

When a standard hammer weight with a diamond tip is made to fall on to the surface of a metal, there will be a great instantaneous load at the point of impact. The kinetic energy of impact is expended in making an indentation at the spot and in rebound of the hammer. If the metal is hard, the indentation made will be small for it yields less, and the height to which the hammer rebounds will be more, and vice versa. If the height of fall of the hammer and its weight are maintained as constant, the height of the rebound of the hammer will give us an idea of the hardness of the metal. Shore's scleroscope operates on this principle.

The instrument essentially consists of a graduated glass tube through which a 2.6 gm. standard hammer falls through a standard height. The whole length of the glass tube is graduated into 100 or 140 equal parts. The height of the rebound may be noted thus:

(i) By watching the top of the hammer itself.

(ii) By means of a magnifying glass and pointer attached to the pipe which are moved to the expected region of rebound. This is done by preliminary trials.

(iii) By noting the number on a circular dial at the top of the instrument.

In all the instruments, the glass tube is aligned to be perfectly vertical. The two important models of Shore's scleroscope, the C_2 and the D are illustrated in Figs. 2.18 and 2.19. In the model D, the circular scale indicates the scleroscope number, in addition to giving conversion into other types of hardness scales.

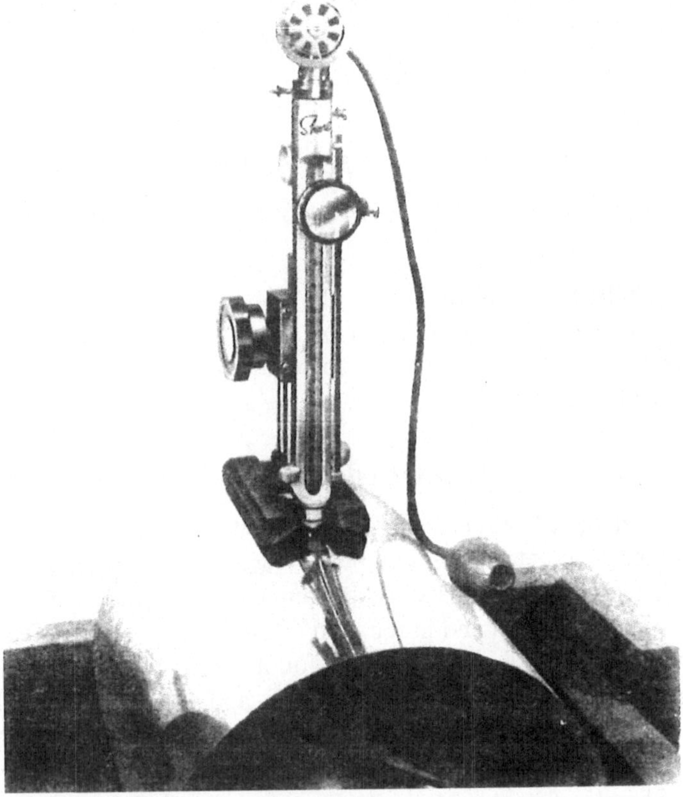

Fig. 2.18 Shore's scleroscope (Model C_2).
(*Courtesy* : The Shore Instrument & Mfg. Company Inc., Jamaica, N.Y., U.S.A.)

Sources of Error

1. The surface condition of the metal under test. If the surface is not perfectly plain, a smaller amount of rebound will occur. Generally it is necessary that the surface of the metal should be more perfect than is necessary in the case of either the Brinell or Rockwell test.
2. The specimen should be perfectly clamped to the base of the instrument, otherwise the rebound will not be proper due to energy losses on account of the movement of the specimen itself. This can be judged by the sound of the impact of the hammer while testing. The sound should be the same as a characteristic hammer blow on metal.
3. Decarburised layers, dust, grease, etc., should not be present on the surface.

Fig. 2.19 Shore's scleroscope (Model D).
(*Courtesy* : The Shore Instrument & Mfg. Company Inc., Jamaica, N.Y., U.S.A.)

Advantages

(i) The procedure is very short and simple.

(ii) The impression left on the object is very small.

(iii) The instrument is portable.

Application

For routine inspection work, especially when 100% testing is necessary, the instrument is invaluable. Very thin sections can also be tested.

Limitations

In this instrument the rebound property of the hard surface is made use of. Since rebound is a characteristic of elastic materials also, elastic materials like rubber, etc., should not be tested by this instrument and the values obtained thereon should not be compared to the hardness of metals. The very nature of the material under test should be similar. Then only can a comparison of the values obtained on testing, be made.

Herberts Pendulum Hardness

This instrument not only determines the hardness of a metal manifested as the resistance to indentation but also as the resistance to work hardening.

In brief, the machine consists of a swinging weight in the shape of bracket. The weight of this pendulum is 4 Kgf. The centre of gravity of the pendulum is designed to act at its top bottom. The pendulum has a length of approximately 30 cm and is capable of accommodating articles of 1,500 mm width or 2,000 mm diameter. A one mm diameter hardened steel ball is fixed in a suitable chuck beneath the pendulum. A skeleton sketch of the instrument is illustrated in Fig. 2.20.

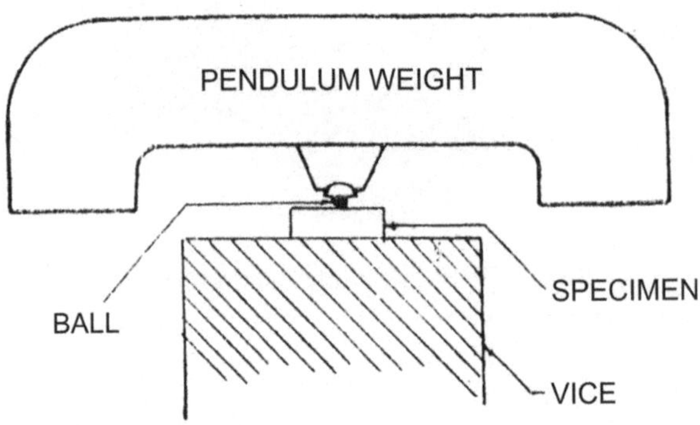

Fig. 2.20 Herbert's pendulum hardness tester.

Time Test

This determines the hardness of the metal. The time required for 5 oscillations of the pendulum is noted and the period is calculated as in the case of the simple pendulum. This is called the time hardness number. There is a factor by which the period can be multiplied to obtain the Brinell hardness number.

The time period will be short if the hardness of the metal is more. When the metal is hard, the indenter ball does not penetrate or get embedded into the metal. Consequently, there will be less resistance to the swinging of the pendulum. On the other hand, when the metal surface is soft, higher values of time period are recorded.

Scale test

The pendulum is tilted through a particular (specified) angle and released and the extent to which it swings on the other side is measured. The angle thus recorded is referred to as the scale hardness number. As explained above the angle of swing depends upon the energy absorbed at the point of contact. Thus, it is a measure of the resistance of the metal to plastic deformation.

From the above two, the 'flow hardness number' is derived, by dividing the scale hardness number by the time hardness number.

The Herbert's Pendulum Hardness (Number) tester is a special purpose machine. Its application is, however, mostly confined to the research laboratory.

A dynamic hardness tester which is very practically usable, even during the processing of a material is the Poldi Hardness tester. The instrument is handy and can be taken to the job when it is unwieldy and cannot be carried to the conventional hardness tester and mounted on its table for testing. It is a simple implement consisting of a hardened steel ball, located firmly in a tubular holder. A movable plunger rigidly held by means of a spring works in the tubular holder. A gap is provided between the ball and the plunger's bottom and into which a 'Standard' rectangular bar is inserted. This bar is hardened to a predetermined value. It has a tapered end to facilitate easy insertion into the space. The standard bar is held in the proper position due to the action of the spring provided in the tubular ball holder (Fig. 2.21).

The surface of the material under test is polished enough to enable the tester to accurately measure the indentation. The process is very simple. The indenter (i.e., the ball holder) is held vertically over the material surface (under test) with the standard bar inserted into it, the ball touching the surface. A hammer blow is given to the plunger from the top. The blow leads to the indentations on the test piece as well as on the standard bar. Both the indentations are measured accurately using a microscope, similar to the one used in Brinell hardness test. The degree of accuracy required is of the order of 0.1 mm.

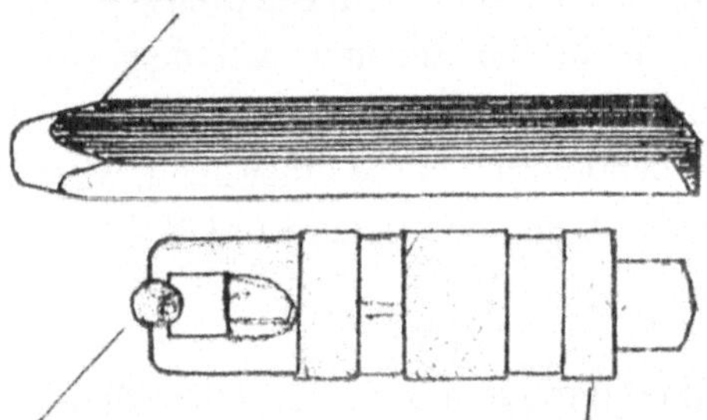

Fig. 2.21

Tensile strength and Brinell hardness number of the material tested can be found out using the standard tables supplied alongwith the instrument. Separate tables are available for steels–annealed state and hardened or hardened and tempered state. Thus, to determine the hardness of the steel, it is necessary that its condition is known. However, even if it is not known, first the tables provided for natural or annealed state are to be used for steels having hardness upto 360. If the hardness exceeds 360, the other tables are to be used.

Separate tables are provided for other metals like copper, aluminium, and alloys like brass, bronze etc. However, these tables do not show the tensile strengths for, the relation between the Brinell hardness number and the tensile strength for these materials is precisely not established. For brass also, two sets of tables are provided, one for castings and the other for rolled brass.

Procedure

The tables consist of vertical and horizontal columns, giving the diameters of indentations on the specimen and those on the standard test bar respectively. Where the horizontal and vertical columns intersect, the corresponding square (in the table) gives the Brinell hardness number and tensile strength of the material.

If a correction factor is given on the standard test bar, the value read from the table is multiplied by the correction factor.

Precautions

1. The hammer used is an ordinary hand hammer of 1/2 Kgf.
2. The blow given should be normal, not very severe nor very mild.
3. The ball holder should be held vertically on the specimen.
4. If the indentation obtained is not circular, the average diameter should be taken to compute the hardness.
5. The standard test bar should be so inserted that the ball rests at a minimum distance of 15 mm for any previous indentation.
6. It should be noted that the Poldi hardness is no substitute for the Bench hardness tester and the values obtained are within 10% of the accurately measured hardness of the material.

HOT HARDNESS

The hardness of metal, just as it is related to the strength at the room temperature, is also found to be related to the high temperature strength at the concerned temperatures (Fig. 2.22). With the advent of the missile era and the prominence given to high temperature materials, the prediction of creep strength may be possible to a certain extent with the knowledge of the high temperature hardness. Methods of high temperature hardness testing are evolved by using a Vickers machine with a sapphire indenter, the indentation being performed in vacuum. There occurs a linear relation between the hot hardness and the strength of the metal; both varying (decreasing) almost parallelly with increasing temperature. High temperature Rockwell hardness testers are also in the market (Fig. 2.23).

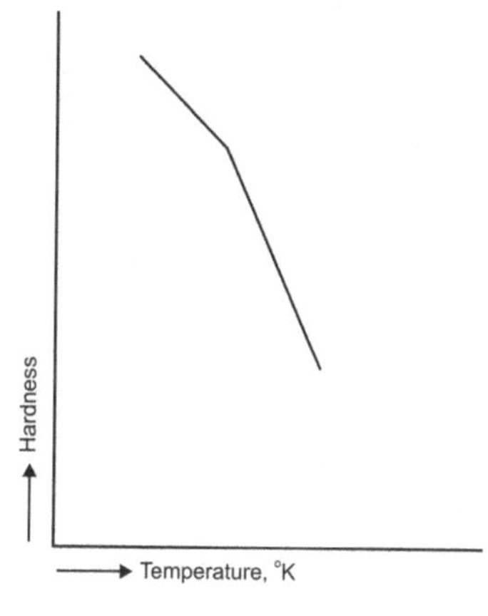

Fig. 2.22 Relation between hardness and high-temperature strength.

Fig. 2.23 High-temperature Rockwell hardness tester.
(*Courtesy* : Wilson Instrument Division of A.C.C.O., Connecticut, U.S.A.)

References

Westerbrook, T. H., *Trans ASM.*, Yol. 45, pp. 221-248, 1953.

Simons. Eric N., *Mechanical Testing of Metallic Materials*, Pitman.

IS: 1500–1968, Method for Brinell hardness test for steel.

IS: 1789–1961, Method for Brinell hardness test for grey cast iron.

IS: 1790–1961, Method for Brinell hardness test for light metals and their alloys.

IS: 3054–1965, Method for Brinell hardness test for copper and copper alloys.

IS: 1586–1968, Method for Rockwell hardness test (Band *c* scales) for steel.

IS: 1501–1968, Vickers hardness test for steel.

IS: 1810–1960, Vickers hardness test for light metals and their alloys.

IS: 2866–1965, Vickers hardness test for copper and copper alloys.

IS: 3754–1967, Calibration of standard blocks to be used for Rockwell B and C scale hardness testing machines.

IS: 4258–1967, Hardness conversion tables for metals.

CHAPTER 3

Impact Tests

GENERAL

Cast iron shows a tensile strength of about 12 tons per square inch. However, if it is hammered with an ordinary hand hammer, it breaks. Based on this we say that cast iron is brittle. There are many other materials which behave like cast iron. Even steel in a totally hardened condition will be brittle. Many a time, an accidental fall of the material from our hands results in its breaking into pieces. This type of brittle behaviour of materials is most undesirable for the common engineering objects.

A simple tensile test does not reveal the brittle nature of the metals, and if only the tensile test data are relied upon and the object put into use, failure is certain. It is, therefore, necessary to test the material under shock or sudden loading conditions.

SIGNIFICANCE

Strength data are mostly academic. In actual use and practice, we rarely come across the ideal case of gradually applied loads. When we sit in a chair, our weight of a few kilograms is applied to the chair, suddenly. When an electric motor is put on, the shaft takes a sudden torque of the full r.p.m., being at zero r.p.m. just before. Many instances can be cited when loads are borne by engineering components suddenly.

IMPACT TESTS

There are four types of impact tests. They are :

(i) Izod,

(ii) Charpy,

(iii) Fremont, and

(iv) Amsler.

Of the above, the first and second are most commonly employed. These machines are standardised in all respects, including the specimens to be used in them.

The principle employed in all impact-testing procedures is that a material absorbs a certain amount of energy before it breaks. The quantity of energy thus absorbed is characteristic of the physical nature of the material. If it is brittle, it breaks more readily, i.e., absorbs a lesser quantity of energy, and if tough, it needs more energy in order to fracture.

The methods of testing are also very similar. A swinging hammer is made to strike the specimen held firmly in a vice. The hammer breaks the specimen on account of its potential energy. The height of rise of the hammer on the other side indicates the residual energy of the hammer. The energy actually absorbed by the material specimen in order to fracture, is given by the difference between initial and final energies of the hammer.

The tests are classed on the basis of the following:

(i) Type of impact used, i.e., the way the impact of the hammer acts to break the material, viz., the tension, torsion, etc.

(ii) The way in which the specimen is mounted, i.e., as a simply supported beam or as a cantilever.

(iii) The type of the specimen used–whether it is a notched specimen or an unnotched specimen.

(iv) The impact load whether applied only once or repeated many times to effect the failure of the specimen.

The Izod Test

The Izod testing machine (Fig. 3.1) consists of a heavy triangular frame. At its apex is a smooth bearing in which swings a heavy pendulum weight. The pendulum weight can be lifted up to a height and clamped. When realeased, it carries an energy of 120 ft. 1bs. (16.36 kg Mts). At the base (illustrated Separately in Fig. 3.2), there is a vice to fix the standard specimen. There is a pointer which moves on a scale at the top to read the energy in Kgf-metres. The pointer is in turn moved by the swinging pendulum weight. The scale is marked from centre to both the ends, so that the energy absorbed by the specimen is read directly.

Fig. 3.1 Izod impact testing machine.
(*Courtesy* : Avery-Denison Ltd., Leeds, U.K.)

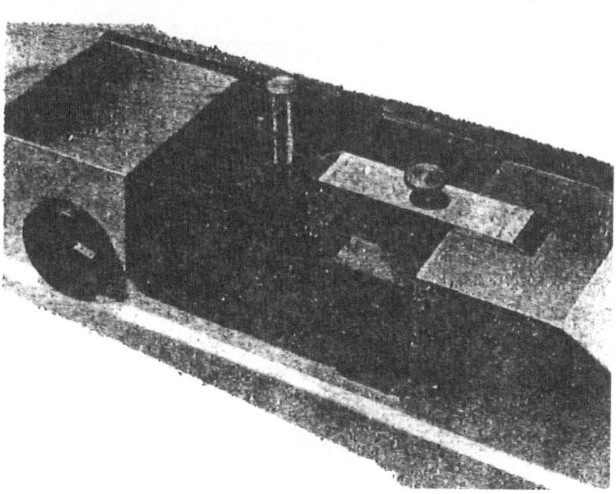

Fig. 3.2 Photograph showing the correct positioning of the specimen for an Izod test.
(*Courtesy* : Avery-Denison Ltd., Leeds, U.K.)

The standard specimen for the Izod test is a square rod of 10 mm side. There is a 2 mm deep, 45° notch made at a distance of 28 mm from the end of the specimen. The root of the notch is finished with a 0.25 mm radius. A cylindrical specimen with a diameter of 11.4 mm and a 3.3 mm deep notch may also be used. The notch employed is also a 45° V-notch, finished with a 0.25 mm root radius. It is customary to take a long test piece, of about 125–130 mm and make two notches on it on the opposite faces or diametrically opposite as the case may be. This serves to perform two tests in order to obtain concurrent values. The specimen is fixed in the machine in such a way that the hammer strikes it at a point 6 mm from the top. The notch of the specimen is fixed to be on level with the anvil, and faces the pendulum. Details of dimensions, etc., are illustrated in Figs. 3.3, 3.4 and 3.5.

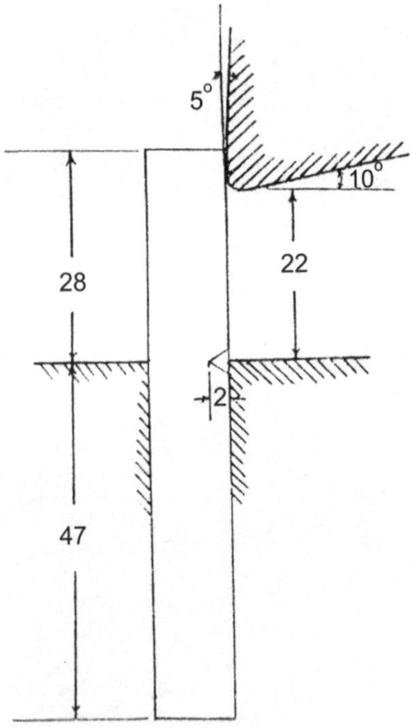

Fig. 3.3 Cantilever beam specimen and mountings for Izod test.

Fig. 3.4 Photograph showing the hammer action in an Izod test.
(*Courtesy* : Avery-Denison Ltd., Leeds, U.K.)

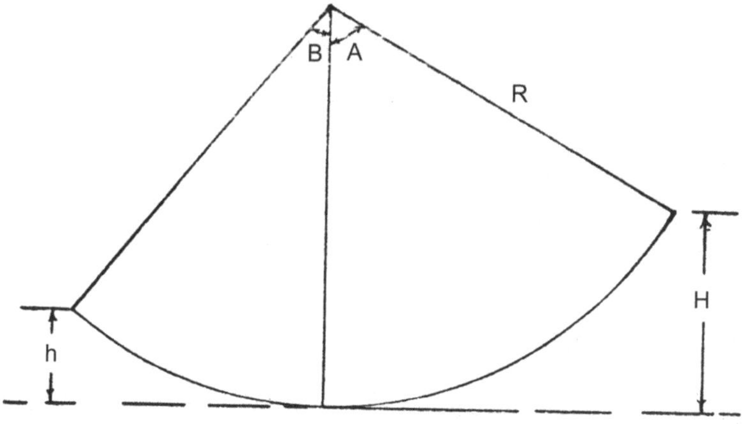

Fig. 3.5 Calibration of Izod pendulum.

The actual testing procedure is as follows:

1. The pendulum weight is brought up and clamped.
2. The specimen is fixed in the vice.
3. The scale is adjusted to read zero.

4. The pendulum weight is released from the clamp.
5. The energy absorbed by the specimen is read on the scale to give the impact strength of the material.

The impact strength, as explained above, is the energy absorbed by the material. It is expressed in ft. lbs., or kilogram force metres.

Corrections and calibration. In the illustration,

Initial energy (the energy in the pendulum before it breaks the specimen) $\left.\right\} = WH = WR\,(1 - \cos A)$

Final energy (the energy remaining in the pendulum after it has broken the specimen) $\left.\right\} = Wh = WR\,(1 - \cos B)$

Energy of rupture of the specimen: $\left.\right\} = W\,(H - h) = WR\,(\cos B - \cos A)$

where

W = weight of the pendulum,

R = length of the pendulum (distance from its centre of gravity to the point of support),

A = angle of fall,

B = angle of rise,

H = height of fall, and

h = height of rise,

A correction to the angle of rise is necessary. The bearing friction is calculated by swinging the pendulum in the air (without any specimen) and noting the pointer readings for a certain number of swings. The decrease in the pointer reading per swing is estimated. This is added to the angle B.

The pendulum not only breaks the specimen but also throws out the fragment. The energy expended in this process should also be taken into account. This should be subtracted from the value of the energy indicated as absorbed by the material in fracture.

If the weight of the fragment be 'w' and 's' the distance of the specimen from the rotational axis of the pendulum, the energy expended in moving the broken fragment will be, $e = w.\,s.\,(1 - \cos B)$.

Representation of the Data

After an Izod test, the following particulars should be furnished, besides the energy absorbed by the specimen:

1. The temperature of the specimen, i.e., the room temperature,
2. model and capacity of the machine,
3. appearance of the fractured surface, and
4. the number of specimens tested.

Izod values for steels

The energy required by the specimen to break (E_T) with a single blow can be considered to be made up of the energy (E_d) to deform the material plastically before fracture and the energy (E_c) to propagate the crack. These energies E_d and E_c are generally not equal, and vary depending upon the nature and condition of the steel. It is noted experimentally that the energy required to fracture a notched specimen is appreciably lower than that required to fracture an unnotched one. Thus, it can be said that the notch lowers the value of E_T. Further, it changes the ratio E_c/E_d.

The relationship between E_c, E_d and E_r are studied by interesting experiments. Steels are broken in two blows instead of a single blow. The magnitude of the first blow is adjusted to be a percentage of its E_T, when a crack should be noticed. In the case of mild steel, a first blow of about 12 % of E_T is found to be sufficient to produce a crack. As the energy of the first blow is increased, it is found that the specimen breaks with decreasing values of blows in the later instances. When a bigger initial blow is given, a larger or deeper crack would be produced. In such a case, the distance through which it should propagate to effect the failure is small. Hence, a smaller second blow is needed subsequently. Hence, it can be said that the ratio E_c/E_T is high for the material.

The case is different with a brittle steel. In such a case, the magnitude of the initial blow has no effect on the energy required to fracture the specimen in the second instance. With the first blow amounting to about 90% E_T, it was found that the specimen either broke off or remained undamaged. This suggests that the initial blows have only deformed the material above the notch elasticity and the striking energy of the first blow is absorbed in the losses and the remaining energy returned to the hammer itself in its rebound. Therefore, the extra energy required for the fracture is the full energy to fracture as if there was no initial blow. Hence the ratio E_c/E_T for the brittle materials is small.

The presence of the notch in the specimen is very important, especially to detect the brittleness. A highly localized stress value is built up at the root of the notch. This causes the commencement of a crack at a much lower value of deformation than for an unnotched specimen. Thus, both the total energy required for the fracture of the specimen E_T, and the relative resistance of the material for the crack propagation E_c / E_T, are of technical and design importance.

Appearance of the fracture

It is advisable that the fracture surfaces of the specimen are also studied besides just recording the energy absorbed in breaking the specimen. The study of the fracture surfaces helps us in determining the impact strength in relation to the transition temperature range of the material. We know that in the transition range, the fracture changes from ductile to brittle. That is, the fibrous fracture with deformation changes into the crystalline fracture. The energy absorbed also decreases. In between these two extremes, the fracture will be a mixed one (Fig. 3.6). It consists of a well-defined crystalline or brittle portion, generally away from the region of the notch.

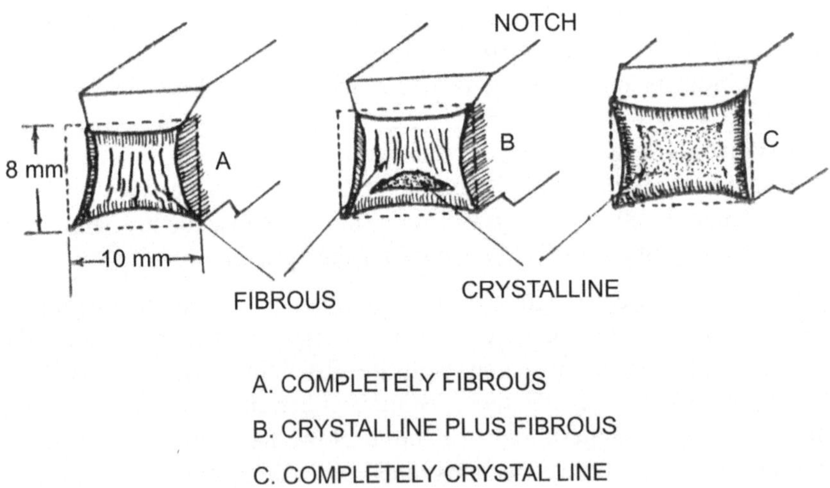

Fig. 3.6 Appearance of impact fracture.

Of course, the extents of the brittle (or crystalline) region and the fibrous (or silky) regions, directly control the impact value. An estimate of the percentage crystallinity computed by the naked eye can be a useful guide and is computed as follows :

$$\text{Percentage Crystallinity} = \frac{\text{Crystalline Area in MM Square}}{80} \times 100.$$

In the case of mild steel the crystalline nature of the fractured area in the region below the crack reveals that the material possesses a high resistance to the initial propagation of the crack.

Generally speaking, the lowest impact values are obtained in the case of fully hardened steels. In these steels, when tempered at a low temperature, say 200°C, the Izod impact value is found to decrease further. On the other hand, a higher temperature of tempering increases the toughness of the steel.

Sl. No.	**Ni-Cr. Steel (Air or Oil Hardened)**	**T.S* Tonnes/cm^2**	**Impact Strength* Kgf. metres.**
1.	Hardened Conditions	15-18	2-3
2.	Tempered at 200°-250° C	15	1.5-2.0
3.	Tempered at 500°-650° C	15	10-12

*Approximate values.

Mild steels and low carbon steels have impact values ranging from 5-12 Kgf. mts.

Plain carbon case hardened steels possess impact values ranging from 6 Kgf. metres onwards.

Charpy Impact Testing Machine

The Charpy impact testing machine differs from the Izod machine in the method of breaking the specimen. The specimen employed may be of any of the three types illustrated in Fig. 3.7. Specimen designs differ only in the shape of the notches. A V-notch as in the case of the Izod specimen, a U-notch, or a keyhole notch are the three commonly recommended for the Charpy specimens.

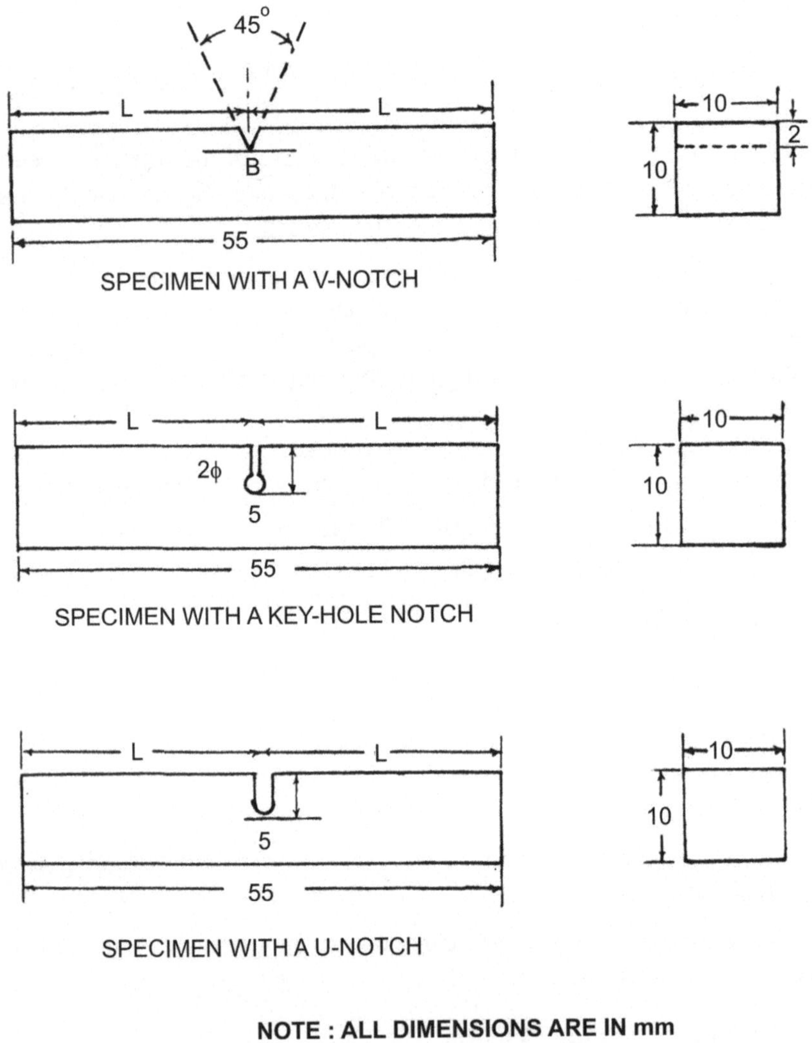

Fig. 3.7 Standard charpy specimens.

The specimen is fixed in the machine as a simple beam (Figs. 3.8 and 3.9). The notch in the specimen does not face the hammer as in the Izod method. The opposite face of the notch is fixed to receive the hammer blow. The hammer head, a pointed one of 8 mm radius, strikes the specimen just in the vertical axis of the notch.

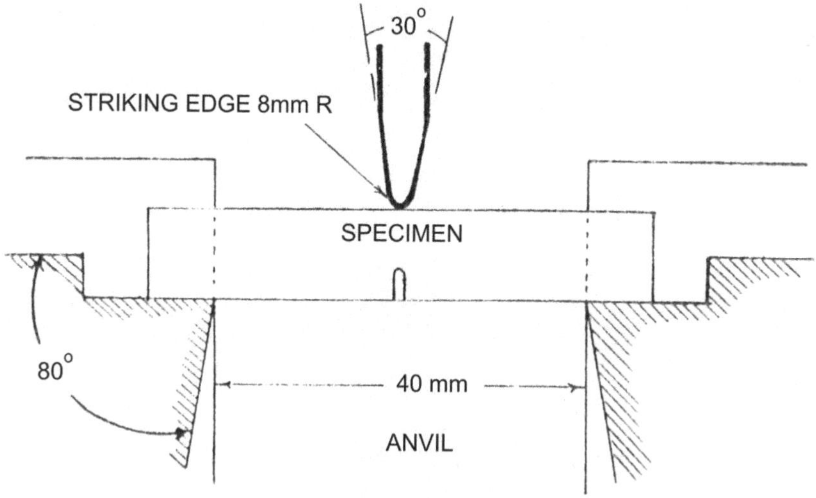

Fig. 3.8 Arrangement of the specimen in the charpy machine.

Fig. 3.9 Photograph showing the correct positioning of the Charpy specimen.
(*Courtesy* : Avery-Denison Ltd., Leeds, U.K.)

The Charpy machine is similar to the Izod machine in the other aspects like the design and method of operation. A hammer blow capacity of 220 ft. lbs. (30 Kgf. metres) is very common. A photograph of the machine is shown in Fig. 3.10.

Fig. 3.10 The Charpy testing machine.
(*Courtesy* : Avery-Denison Ltd., Leeds, U.K.)

Tensile impact test

The Charpy machine is also used for determining the tensile impact strength of the materials. Either a specimen with a circular section with a 6 mm diameter is used, or one having a diameter of 10 mm with a circular notch of 1 mm width and 2 mm depth is used (Fig. 3.11). It is interesting to note here that the net diameter of both the specimens is 6 mm.

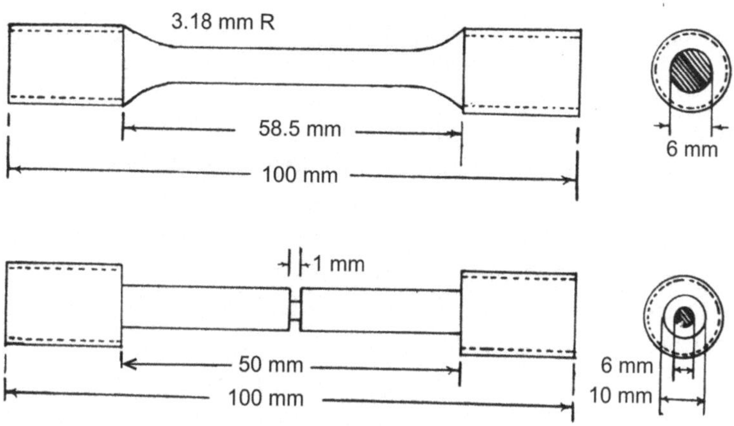

Fig. 3.11 Tensile impact specimens.
(From A.S.M. Metals Handbook, 1948).

The specimen is fixed to the rear end of the pendulum hammer. A dead weight, to increase its net width, is secured to the far end of the specimen. The anvils, which are used to fix the specimen is a simply supported beam, are brought together (Fig. 3.12). This reduces the space in between them and permits only the specimen to pass through. Consequently, the far end of the specimen is suddenly held back by the anvils. Thus a tensile stress is developed in the specimen under conditions of impact and the specimen is broken.

Fig. 3.12 Photograph showing the tensile-impact specimen fixed in position for the tensile impact test.
(*Courtesy* : Avery-Denison Ltd., Leeds, U.K.)

The tensile impact test is yet to be standardized. Its application is also very limited.

The ordinary Charpy test measures the total energy absorbed in fracturing the specimen. Additional information regarding the test can be obtained by suitable instrumenting the machine to provide the load time history of the specimen during the test. It is possible to determine the energy required for initiating the fracture and the energy required for propagating it.

Data acquisition software is a software for material testing designed to acquire, analyse and store test data. One such software developed is Dynatup 930-I with which the mechanical properties of most of the materials including paper, textiles, rubber and plastics in accordance with national and international standards.

The software enables us to create our own test set-up, customize test reports and export the files to database programs. The software features Microsoft windows-based graphical user interface which facilitates to quickly and efficiently get graphic displays of load, energy, velocity, time or deflection within seconds after the test. Test results can be saved for later review and analysis. Multiple tests can be simultaneously viewed using the Batchset feature.

The operation of 930-I Data acquisition software in its simplest form is very easy when applied to impact testing. Whether it be Instron or Dynatup setup, basically consists of a load transducer or Tup which measures the force resulting from the pendulum's contact with the specimen. The data acquisition system software takes the Tup signal and converts it into a usable signal.

There are two types of Tups, one a Semiconductor strain gauge Tup and the other Piezoelectric Tup. The higher frequency Piezo Tups are ideally suited for testing their films and other ultra-low load materials. Tups or strain gauges are high precision load transducers. With use, their sensitively may change and hence it is recommended to get the Tup calibrated on a biannual basis.

The velocity detector has a two-fold use. It triggers a signal used for starting data acquisition and used to determine the impact velocity.

The signal-conditioning unit is a stand-alone unit that receives the impact signals from the impact-testing machine and provides processing for both the signals before passing them on to the A/D board in the computer. The S.C. unit provides filtering and amplification of the Tup signal and calculates the impact velocity based on the pulsed timing signal received from the velocity detector. A graph received from the computer is shown in Fig. 3.13, which shows how elaborate data can be obtained with this computerized attachment.

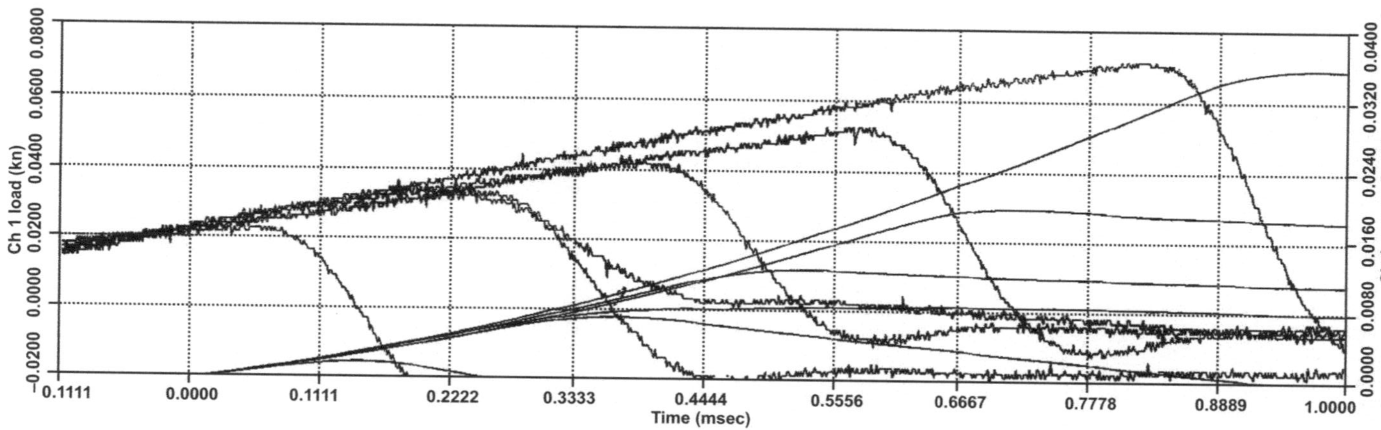

Impact velocity (l) (m/sec)	Total deflection (l) (mm)	Time no	Impact energy (l) (joule)	Impact velocity (l) (m/sec)	Maximum load (l) (kn)	Energy to max load (l) (joule)	Time to max load (l) (msec)	Defl at max load (l) (mm)	Total energy (l) (joule)
0.7702	0.5496	1 - ■	8.0744	0.7702	0.0529	0.0154	0.5490	0.4241	0.0199
0.7702	0.2818	2 - ■	8.0740	0.7702	0.0349	0.0052	0.2410	0.1867	0.0073
0.7699	0.7466	3 - ■	8.0663	0.7699	0.0722	0.0298	0.8030	0.6185	0.0358
0.7702	0.4093	4 - ■	8.0744	0.7702	0.0433	0.0095	0.3730	0.2883	0.0128
0.7703	0.1118	5 - ■	8.0756	0.7703	0.0241	0.0000	0.0010	0.0019	0.0020
0.7703	0.4740	6 - ■	8.0762	0.7703	0.0373	0.0039	0.1800	0.1396	0.0086
0.7702	0.4289	Average	8.0735	0.7702	0.0441	0.0106	0.3578	0.2765	0.0144
0.0003	0.2191	StdDev	0.0036	0.0003	0.0167	0.0108	0.2855	0.2196	0.0121

Fig. 3.13

EFFECT OF VARIABLES ON THE IMPACT TEST VALUES

Various impact test data do not agree. The variations are due to the following :

(i) A part of the energy of the pendulum being absorbed by the machine and its foundation, instead of the specimen alone.

(ii) Variations in the striking velocity of the pendulum.

(iii) Size and shape of the specimen.

(iv) Temperature.

(v) The grain size.

(vi) Composition of the metal.

(vii) Cold work.

A. Energy Losses

Theoretically, all the energy of the swinging hammer should be transferred to the specimen. In practice, this does not happen. When the pendulum hammer comes and strikes the specimen, the vice and, thereby, the foundation of the machine take up some of the energy of the pendulum. However, this is minimized by the manufacturers of the machines by providing a heavy structure for the machine. Even while erecting the machine, a very thorough foundation work is laid to minimize the damping and vibrations.

B. Striking Velocity

The commonly used Izod and Charpy impact testing machines develop 16.6 and 36.6 Kgf metres energy respectively; their respective striking velocities being 3.5 and 5.3 metres per second.

Changes in the velocity of impact in these ranges are found to have no effect on the impact strength values obtained. However, at very high velocities of impact, the impact resistances are found to decrease remarkably. It is pointed out that, for a particular metal, there is a critical velocity of impact above which it gives way very easily. Further, the micro-structural condition is also found to have its influence on the critical striking velocity; an annealed steel having a much lower critical striking velocity than a hardened steel.

C. Size and Shape of the Specimen

Photographical decrease in the size of the specimen may be and is sometimes adopted when sufficient material to be tested is not available to prepare a specimen of the standard size. It is experimentally established that the impact strength value is not effected by the size of the specimen except when the steel is brittle. Brittle steels, however, record a higher value of impact when tested on miniature sizes. The same is found to be true when the testing is done at lower temperatures.

The presence of the notch and its sharpness have a marked effect on the impact values. This aspect is covered in the next chapter while discussing the energy absorbed by the specimen in order to develop a crack and to propagate an already developed crack to failure.

I.S.O. recommended tolerance on the standard test specimen dimensions are ± 0.11 mm. Malmberg has shown that these tolerances are all right but for the depth of the notch. Hence, it is advisable that the depth of 2 mm of the notch be carefully, rather stringently, adhered to.

It is observed that the behaviour of a metal is more sensitive to the presence of the notch than its size. That is why, the shallow V-notch is preferred to the keyhole notch in the Charpy impact test.

D. Temperature

Impact strength of notched bars is greatly affected by the temperature. Up to a certain temperature, the metal exhibits a brittle fracture. After that temperature is exceeded and until a higher temperature is reached, the nature of fracture will be a mixed one, viz., ductile and brittle. Again, after passing this range, the metal shows a ductile fracture (Fig. 3.14).

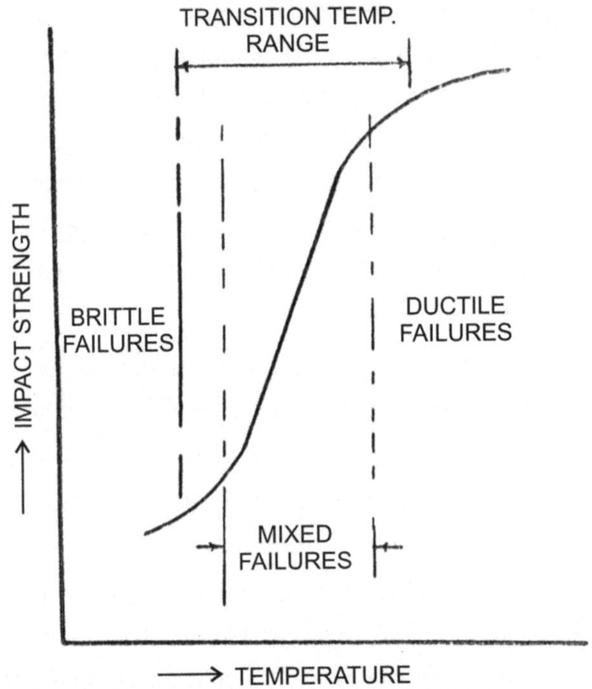

Fig. 3.14 Impact strength-temperature relationship.

It can be noted that the change of fracture from brittle to ductile is a gradual process and occurs over a range of temperature. This is called the transition temperature range. Further, it should be noted that the metal breaks very easily below the transition temperature range (brittle failure). The embrittlement phenomenon is studied keeping this feature in view.

E. Grain Size

The effect of grain size on transition temperature is quite marked. As the grain size decreases (the A.S.T.M. grain size number increases) the transition temperature is found to be lowered considerably. Thus, it is beneficial to either adopt the lowest possible finishing temperature in hot working or normalize the steel after hot working.

F. Composition

Of all the alloying elements in steel, chromium has little effect on the transition temperature while the maximum effects are produced by carbon and manganese. While carbon pronouncedly raises the transition temperature, manganese, though to a slightly lesser degree, lowers it. Their effects are such that for a satisfactory notch toughness, they should be present in a proportion Mn: C = 3: 1.

Other elements which raise the transition temperature are phosphorus and to a lesser degree, silicon, when present in quantities more than 0.25%. Of course, the effect of molybdenum on the transition temperature is as drastic as that of carbon, but fortunately all the steels do not contain molybdenum unless it is intentionally added.

Oxygen is another element which raises the transition temperature considerably. Every thousandth percentage of it (0.001%) raises the transition temperature by about 5° C (on a 0° C basis). Thus, the process of killing the steel has a bearing on its impact properties. Killed steels exhibit better impact properties than the semi-killed and rimming steels. Again amongst the killed ones, aluminium-killed steels are found to be tougher than the silicon-killed steels. Aluminium forms the nitride also, besides oxide, and thus seals off the nitrogen. Nitrogen-embrittlement is thus avoided.

Besides manganese, which lowers the transition temperature, nickel appears to be the only alloying element which lowers the transition temperature. Nickel in quantities of about 2% is very useful in this regard.

Microstructure and heat treatment

Completely pearlitic structure imparts very poor impact properties to the steel, while a structure consisting of tempered martensite yields good notch impact strength values. Bainitic structure is of a moderate quality in this respect.

Embrittlement phenomena

Temper embrittlement and hydrogen embrittlement are discussed below.

Temper embrittlement

Steels are found to lose toughness remarkably when held between 350°C and the transition temperature. This is referred to as the temper embrittlement. Tempering or slow cooling in the aforesaid temperature range should be avoided to get rid of this phenomenon.

While heat-treating heavy sections, it becomes practically impossible to avoid the embrittlement temperature range. Addition of molybdenum is found to reduce the susceptibility to temper embrittlement.

Whether a steel is prone to temper embrittlement or not can be found out by determining its transition temperature. The transition temperatures of temper brittle steels are observed to be about 100°C or so above those for ordinary steels. Further, fracture surface shows out to be intercrystalline, unlike the transcrystalline nature of an ordinary brittle failure.

It is established that temper embrittlement results on account of grain boundary weakness. Though no experimental evidence is available, it is assumed that a precipitate or film of an unobservable fineness is formed at the grain boundaries.

Temper brittleness may be avoided by:

(i) Tempering at a higher temperature for a shorter time than vice versa,

(ii) alloying with molybdenum, and/or

(iii) quenching in water from the tempering temperature.

Hydrogen embrittlement : When hydrogen is present in a metal or alloy, it produces severe embrittlement. Quantities of hydrogen as small as 0.0001% cause the steel to crack. It is observed that B.C.C. and H.C.P. metals are highly susceptible to this effect.

Hydrogen embrittlement is affected by strain rate and temperature. The chief evidence for this phenomenon is a delayed fracture. Slow tension and bending tests detect the hydrogen embrittlement during which the ductility will be found to be very low. Impact tests do not reveal this effect. Steels are not prone to hydrogen embrittlement both at the subzero temperatures and high temperatures. It is very marked at the room temperature.

While testing for hydrogen embrittlement, what is called a delayed fracture curve is plotted, with the stress as the ordinates and the time as the abscissa (Fig. 3.15). The curve resembles a typical *SN*-curve. Hence this phenomenon is often called “static fatigue”. The minimum stress below which the fracture will not occur is referred to as the “static fatigue limit”.

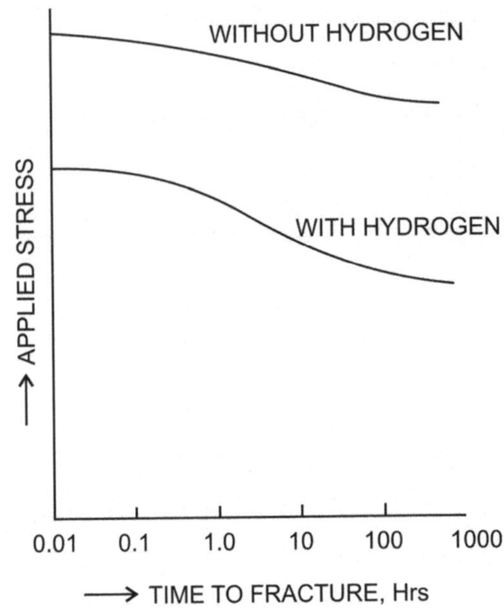

Fig. 3.15 Delayed fracture curve.

Hydrogen dissolved in the steel is present in the atomic state. Under favourable conditions, however, these atoms diffuse into any voids or regions of high dislocation concentration and become molecular. High pressures develop and start to exert, which ultimately lead to the crack formation and propagation.

The dissolved hydrogen in the metal can be driven off to a considerable extent by prolonged heating at 200-250°C.

G. Cold Work

The same metal in a cold worked condition and in an annealed condition exhibits two types of impact strengths. Work-hardened metals show brittle failures and annealed ones ductile failures. The reason for the brittle failure in the case of cold worked metal may be the orientation of the grains in the direction of work (preferred orientation) and the consequent behaviour of the metal as if it were a fibrous one (isotropy).

CALIBRATION

According to the I.S.O. recommendations, the machine scales can be calibrated using unnotched test pieces of steel the widths of which are kept constant and thicknesses varied in a controlled fashion, the impact values of which are known before. It is recommended by the I.S.I. that the pendulum impact testing machines are better calibrated once in three years. The precise method of calibration is out of the scope of this book.

References

A.S. Tetelman and A.J. Mc Evily Jr. *Fracture of structural Materials,* John wiley and Sons. Inc, Newyork. 1967.

Tipper, C.F., *The Brittle Fracture Story*, Cambridge University Publication, 1962.

Symposium on impact testing, *A.S.T.M. Sp. Tech. Publication*, 1962.

Parker, E.R., *Brittle Behaviour of Engineering Structures*, John Wiley, 1957.

IS: 1499-1959, Charpy impact test for steel.

IS: 1598-1960, Method for Izod impact test for steel.

IS: 5070-1969, Unnotched impact test for grey cast iron.

CHAPTER 4

Fracture Mechanism

When a material is loaded and the stress reaches a value that the material cannot withstand, it breaks. In material testing terminology this is known as fracture. Fracture can be defined as the separation of a material into two or more pieces under stress.

A fundamental study of the process or mechanism of fracture can be made by observing the two fracture surfaces. How the material has ruptured, gives us an idea regarding the behaviour of the material during and prior to the rupture. A brittle material shows no visible plastic deformation at the time of its fracture (Fig. 4.1). The two fractured pieces when joined together will exactly resemble the test specimen before loading. This is not common in metals and alloys but occurs in non-metals and viscous materials like glass.

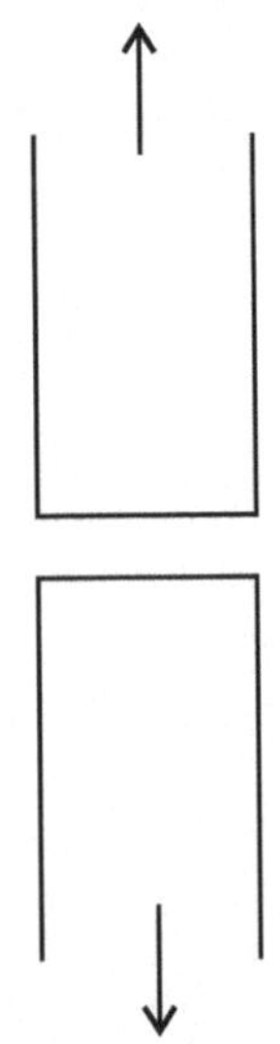

Fig. 4.1 Brittle failure.

Metals and alloys show a different type of behavior. A degree of plastic deformation can be found during or prior to the failure. A drastic case is complete necking or drawing-down to too nearly conical surfaces. This is exhibited by very ductile materials like annealed copper. A mixed type of fracture commonly known as the 'cup and cone fracture' is exhibited by a majority of metals and alloys. The cup and cone fracture consists of a core of plastic deformation and an annular ring of brittle failure (Fig. 4.2). Whenever we refer to a ductile failure, it means without saying the cup and cone type of fracture only, as it is the most common in metals.

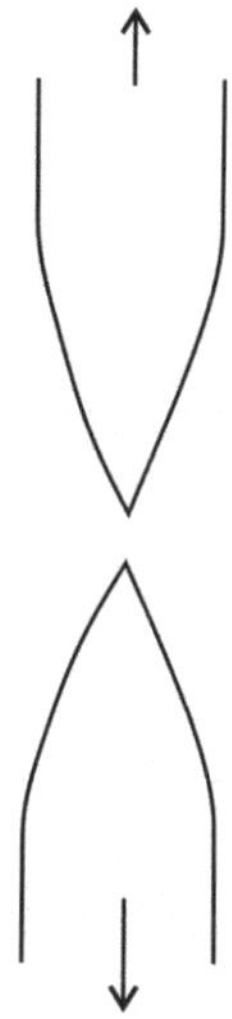

Fig. 4.2 Ductile fracture.

The cup and cone fracture results by a gradual process. The specimen, under the loading conditions prevailing, necks down at the yield stress. When once the necking occurs, the stress distribution which had thereto been uniaxial, changes into a three-dimensional pattern. Small, weaker regions are supposed to be nucleated at this stage and propagate into a minute crack under the influence of the multiple stresses. This crack, generally perpendicular to the direction of the loading, propagates. The crack propagation is governed by the ductility of the material and failure occurs progressively by the plastic deformation reaching its peak in the localized sites. Each region of protrusion (visible in the fracture) can be linked to the complete drawn-down state in a completely ductile fracture

(Fig. 4.3). The crack propagates in this way to the vicinity of the necked surface. Thereafter, the failure will be by cleavage, generally at 45° to the vertical axis. This cleavage region is the brittle fracture region.

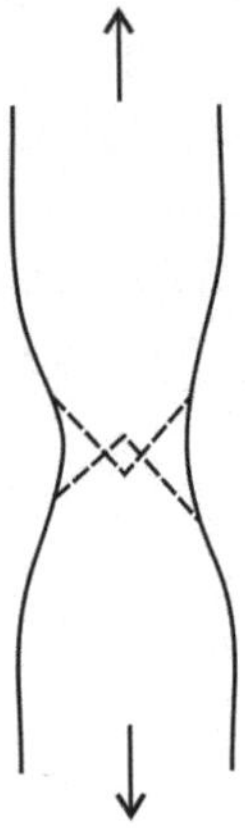

Fig. 4.3 Cup and cone fracture.

COHESIVE STRENGTH

The cohesive strength of metals is very high. A concept of the cohesive strength can be had from the interatomic binding of the solids. Cohesive forces, i.e., the resultant attraction between the atoms, is the difference between the attractive forces between the nuclei of the atoms and the electrons, and the repulsion between the overlapping electron zones. A minima occurs in the energy curve for a particular distance, the interatomic spacing between the atoms. Thus, most of the metals are 'rigid solids'.

If the cohesive strength σ can be plotted against the distance between the atoms, a curve, which can be approximated to a sine curve, is obtained (Fig. 4.4). A tensile load applied tends to increase the separation between the atoms. When plastic deformation starts, the distance between the atoms is increased. Under such conditions, the rate of decrease of the repulsive forces is more than the net forces of balance between the atoms. A condition when the repulsive forces become negligible and the attractive forces also decrease due to the increased separation of the atoms is reached. This corresponds to the peak of the curve, σ max., the cohesive strength of the body.

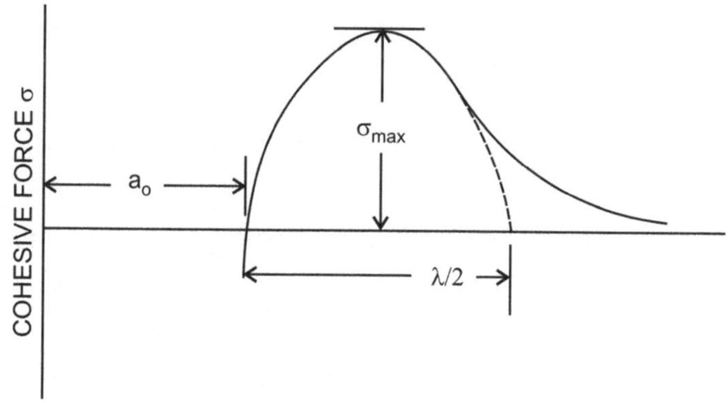

Fig. 4.4 Cohesive force *vs* separation between atoms.

Assuming the curve to be sinusoidal, we get the expression for the curve as,

$$\sigma = \sigma_{max}.\ \text{Sin}\ \frac{2\pi x}{\lambda} \quad ...(4.1)$$

The work done can be calculated by the area under the curve which can be obtained by integrating the eq. (4.1) between a_o and $(a_o+\lambda/2)$ thus,

$$\text{Work done} = \int_{a_o}^{(\lambda/2+a_o)} \sigma_{max.}\ \text{Sin}\ \frac{2\pi x}{\lambda}\ dx. = \frac{\lambda.\ \sigma_{max.}}{\pi} \frac{\cos\ 2\pi}{\lambda},\ a_o \quad .....(4.2)$$

If we imagine that the work done is completely utilized in the formation of a crack and if γ represents the energy required to create a new surface, then we have

$$\frac{\lambda.\ \sigma_{max.}}{\pi} = 2\gamma \quad \text{or} \quad \sigma_{max} = \frac{2\pi\gamma}{\lambda} \quad ...(4.3)$$

Now we concentrate in disposing of the constants π and λ. For this purpose we approximate the ascending portion of the sine curve (to the tune of λ/4) to be a straight line. Then, Hooke's law can be applied and can be expressed as,

$$\sigma = \frac{E.x}{a_o} \quad ...(4.4)$$

where E is the modulus of Elasticity of the material. Differentiating eq. (4.1) with respect to x we get,

$$\frac{d\sigma}{dx} = \sigma_{max}.\ \frac{2\pi}{\lambda}.\ \text{Cos}\ \frac{2\pi x}{\lambda}$$

As cos $2\pi x/\lambda$ can be taken as unity since the practical deformations are small (x is small), we can omit the term in the above expression, thus,

$$\frac{d\sigma}{dx} = \sigma_{max}.\ \frac{2\pi}{\lambda} \qquad .....(4.5)$$

Again, differentiating the eq. (4.4) with respect to x,

$$\frac{d\sigma}{dx} = \frac{E}{a_o} \qquad .....(4.6)$$

Thus,

$$\frac{d\sigma}{dx} = \sigma_{max}.\ \frac{2\pi}{\lambda} = \frac{E}{a_o}$$

$$\frac{2\pi}{\lambda} = \frac{E}{a_o.\ \sigma_{max}.}$$

Substituting for $2\pi/\lambda$ in eq. (4.3) we have

$$\sigma_{max} = \left[\frac{E\gamma}{a_o}\right]^{1/2} \qquad .....(4.7)$$

This expression is very handy. The values of strength computed from the above equation (theoretical strength) will be many times more than the observed values. However, it is observed that we can go nearer the calculated values by experimenting on single crystal whiskers. This revelation emphasizes the effect of dislocations on the properties of crystal aggregates. Let us examine this in some detail the crystal structure and then crystal behavior under load.

All materials consist of basically atoms. These atoms are arranged in a regular fashion. This arrangement is called a Space Lattice (Fig. 4.5). In this arrangement atoms occupy sites, which are regular and predictable. In other words, the distance between two atoms is equal in all directions. The minimum number of atoms taken as a shape (e.g., a cube), which repeats itself in all the three directions in the lattice is called a 'unit cell'. The shape of the unit cell defines the atomic structure. If the unit cell is a cube, structure of the lattice is defined as cubic and if it is rhombic, the structure is referred to as rhombic and so on.

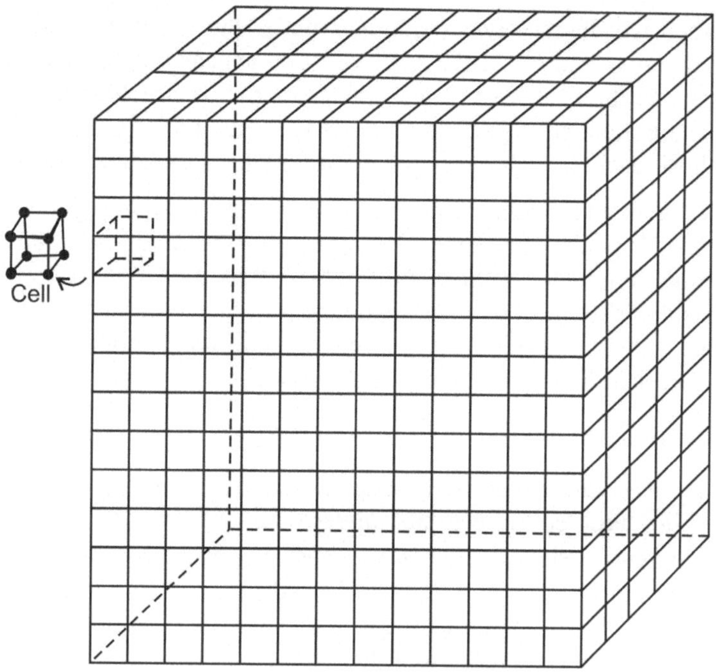

Fig. 4.5 Space Lattice.

The atoms do not occur in a periodic fashion over long distances. Their direction changes after some distance. This changed directions continues for a certain distance and again the direction alters. The existence in a particular direction is called orientation. The region of uniformity is called a crystal and the altered direction in the adjacent region is another crystal. Oftentimes, the changes in the orientations are not abrupt but gradual (over atomic distances). These regions between two crystals are called grain boundaries. This crystalline nature of metals is not visible to naked eye. The metal surface should be polished to a mirror finish and suitably etched with very weak acid solution and observed under a microscope at a magnification of at least 100. The study of metals and alloys under microscopes is known as metallography. Grain (crystal) structures are revealed under a microscope. The regions between two grains, is revealed as lines separating the grains and these are called grain boundaries.

When we observe two specimens under the microscope at the same magnification and when the grains in one appear bigger than the other, we say that the former metal has a coarse grained structure. The structure with more number of grains under view is called *fine grained.* Reporting like this is arbitrary. There are prescribed methods to define the grain size.

As seen above, the lattice orientation in adjoining grains will be different. It is also noted that the change in orientation is gradual. Thus, the grain boundary region will not contain atoms in any stipulated (definable) pattern. These grain boundaries are regions of disturbance. Further, any impurities in the metal will concentrate at the grain boundaries adding to the atomic disturbance.

Grain Structure

So far, we studied about the space lattice taking the example of a cubic cell. The 'unit cell' need not always be cubic. It is defined by its parameters–its edges a, b, and c and angles α (between b and c), β (between a and c) and γ (between a and b). Fig. 4.6 serves to visualize this better. There are fourteen types of space lattices and they fall into seven crystal systems listed in Table. 4.1.

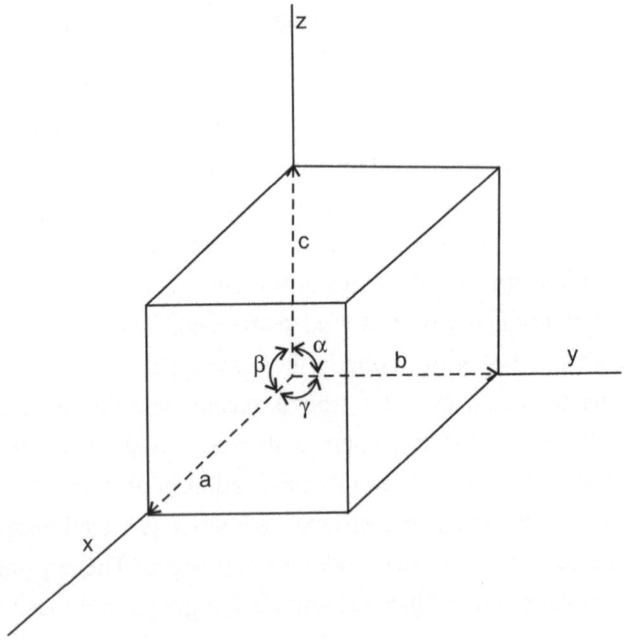

Fig. 4.6 Lattice Parameters.

Table 4.1 Crystal systems*

S.No.	Lattice	Description
1.	Triclinic	Three unequal axes, no two of which are perpendicular $a \neq b \neq c$ $\alpha \neq \beta \neq \gamma \neq 90^\circ$
2.	Monoclinic	Three unequal axes, one of which is perpendicular to the other two $a \neq b \neq c$ $\alpha = \gamma = 90^\circ \neq \beta$
3.	Orthorhombic	Three unequal axes, all perpendicular $a \neq b \neq c$ $\alpha = \beta = \gamma = 90^\circ$
4.	Rhombohedral (trigonal)	Three equal axes, not at right angles $a = b = c$ $\alpha = \beta = \gamma \neq 90^\circ$
5.	Hexagonal	Three equal coplanar axes at 120° and a fourth unequal axis perpendicular to their plane $a = b \neq c$ $\alpha = \beta = 90^\circ$ $\gamma = 120^\circ$
6.	Tetragonal	Three perpendicular axes, only two equal $a = b \neq c$ $\alpha = \beta = \gamma = 90^\circ$
7.	Cubic	Three equal axes, mutually perpendicular $a = b = c$ $\alpha = \beta = \gamma = 90^\circ$

*From C.S. Barrett, "structure of Metals", McGraw-Hill Book Company, Inc., New York, 1952.

Again it is our fortune that almost all the metals crystallize into three important crystal structures. These are Face Centered Cubic (FCC), Body Centered Cubic (BCC) and Hexagonal Close Packed (HCP). These are illustrated in Figures 4.7, 4.8 and 4.9.

Face Centered Cubic (FCC)

This structure consists of a cube with eight atoms situated at the eight corners (Fig. 4.7). It also consists of an atom at the centers of each of the six faces of the cube (and hence the name). In this structure, each of the

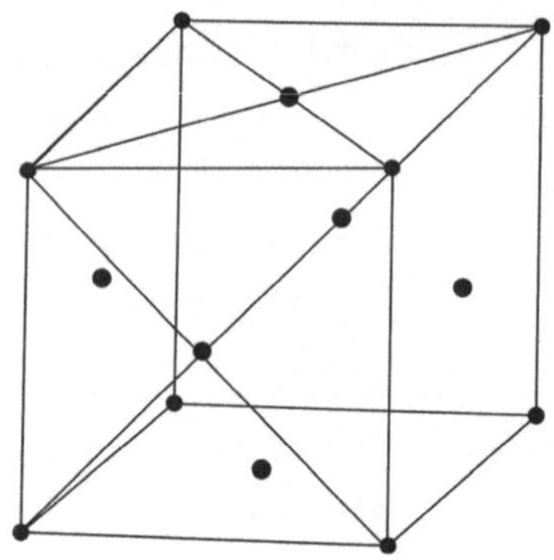

Fig. 4.7 F.C.C Lattice.

corner atoms shared by eight cubes. Each of the face centered atoms shared by two cubes. So we can calculate the atomic density of the structure thus,

In one lattice, 8 corner atoms ----------- $8 \times \frac{1}{8} = 1$

In one lattice, 6 face center atoms ----------- $6 \times \frac{1}{2} = 3$

Total = 4 atoms

The unit cell of FCC structure consists of 4 atoms.

Body Centered Cubic (BCC)

This structure consists of a cube with eight atoms situated at its eight corners (Fig. 4.8). There also exists an atom at the center of the cube. Thus,

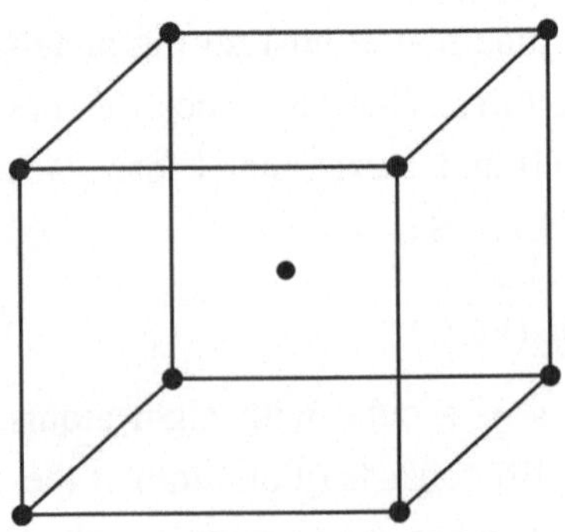

Fig. 4.8 B.C.C. Lattice.

In one lattice, 8 corner atoms ----------- $8 \times \frac{1}{8} = 1$

In one lattice, 1 central atom ----------- $= 1$

Total = 2 atoms

The unit cell of BCC structure contains 2 atoms.

It can be seen from the above that the FCC cell is more densely packed than a BCC cell. Metals like Chromium, Tungsten, Molybdenum, Vanadium, Sodium and α and δ irons possess this structure. Metals like Aluminium, Copper, Gold, Silver, Lead and γ iron are some examples possessing FCC structure.

Hexagonal Close Packed (HCP)

A hexagonal close packed lattice shows two basal planes which are regular hexagons containing an atom at each of the hexagon corners and one at their geometrical centers (Fig. 4.9). In addition to these atoms, each hexagon contains three atoms at the corners of a triangle in the center of the prism. Their location will be inside the hexagon and inside the alternate rectangular sides.

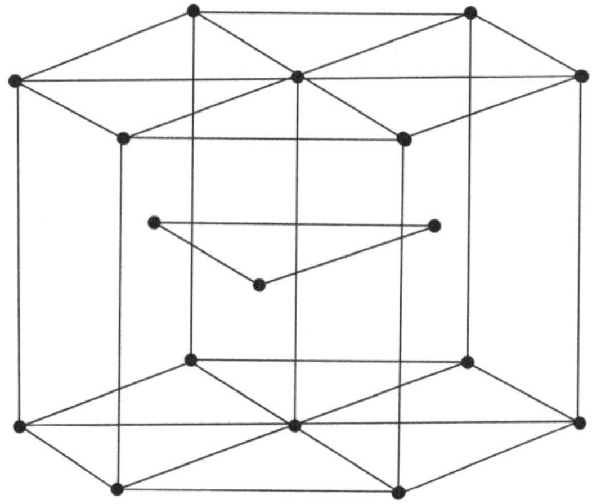

Fig. 4.9 H.C.P. Lattice.

Since each corner atom of the unit cell is shared by eight cells, the basal atoms shared by two and the three atoms inside the hexagon are exclusively of the lattice, the atomic density of the lattice is computed as,

12 corner atoms	---	$12 \times \frac{1}{6}$	$= 2$
2 basal atoms	---	$2 \times \frac{1}{2}$	$= 1$
3 central atoms	---		$= 3$
Total			$= 6$

Examples of metals that crystallize in this type of structure are magnesium, beryllium, zinc, cadmium etc.

The unit cell of a cubic system can be specified by a single lattice parameter i.e., the side of the cube, a. A hexagonal cell requires the width of the hexagon, a and the height (the distance between the two basal planes), c. The axial ratio $\frac{c}{a}$ which is sometimes given. The axial ratio for beryllium is given as 1.58 and for cadmium 1.88.

Polymorphism and allotropy

Polymorphism is the property of the material to exist in more than one type of crystal lattice in solid state. If the change (of structure) is reversible, the polymorphic change is called allotropy. About a dozen metals are known to show this allotropic changes, but the most important and best known is iron. Iron solidifies at 1532° C as δ–iron BC C, while cooling further, the δ–iron changes to γ–iron at 1400° C which again becomes α-iron at 910° C. The γ-iron has an FCC structure while the α-iron has again BCC structure.

Crystal Defects

It is possible to calculate the theoretical strength of a metal by the force required to separate the bond between adjacent atoms. This turns out to be several millions kgf/cm^2. Ordinarily the strength of metals is 100 to 1000 times less. The reason for this lies in the occurrence of defects in the crystal structures of metals.

Real crystals do not contain the atoms in perfect symmetry. Some regions will have defects or imperfections. When the deviation from the periodic arrangement of the atoms of the lattice is localized to the vicinity of a few atoms, the defect is called a point defect or point imperfection. If this extends through a microscopic region of the crystal, it is called Lattice imperfection. The lattice imperfections can again be divided into line defects and surface or plane defects.

Point Defects

These are mainly of three types–the vacancy, the substitutional atoms and the interstitial atom. When an atom is found missing from a normal lattice position, a vacancy exists (Fig. 4.10). A number of vacancies can be

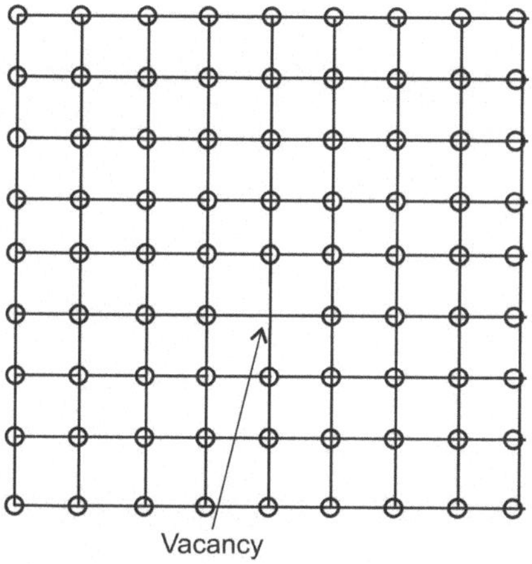

Fig. 4.10 Vacancy.

caused in a pure metal by thermal excitation and these are thermodynamically stable at all temperatures above absolute zero.

$$\frac{n}{N} = e^{\frac{-Es}{KT}}$$

Where,

n = No. of vacancies at the absolute temperature T

N = Total no. of lattice sites of atoms.

Es = Energy required to move an atom from interior to the surface

K = Boltzman's constant

An atom that is trapped inside a crystal lattice at a point intermediate between normal lattice positions is called an Interstitial atom (Fig. 4.11). The presence of an impurity atom at a lattice position is when there occurs a vacancy or the energy requirements are fulfilled. This is called a substitutional atom (Fig. 4.12). Both interstitial and impure atoms result in disturbance of the periodicity of the lattice.

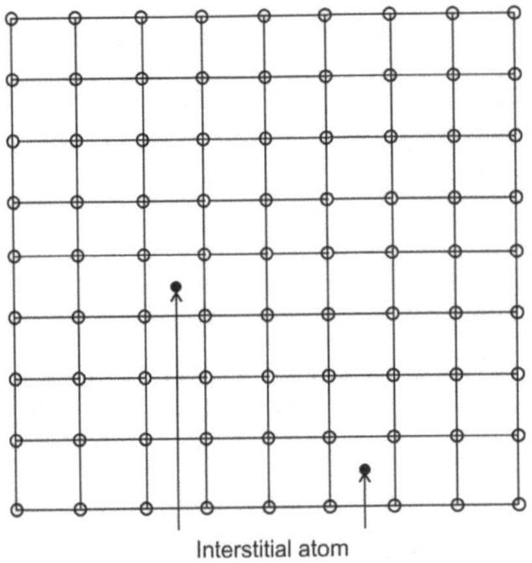

Fig. 4.11 Interstitial atom.

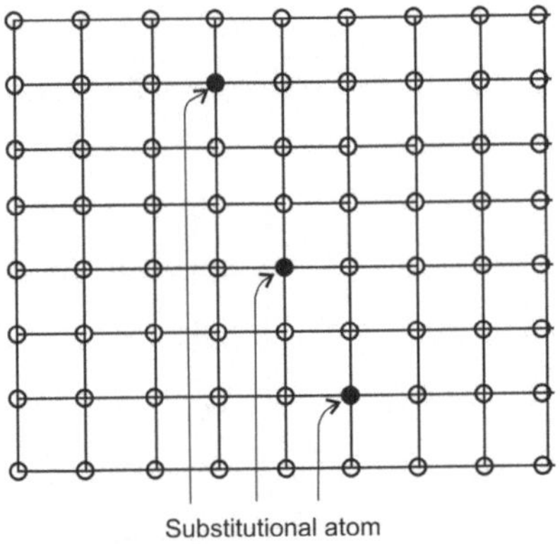

Fig. 4.12 Substitutional Atom

Line Defects/Dislocations

Dislocation is the most important two-dimensional line defect (Fig. 4.13). This can be imagined to be the region of disturbance between two, otherwise perfect lattice structures. This can also be said as the boundary between the slipped and unslipped regions of a crystal. The discontinuation of a row of atoms (in section) is called an edge dislocation. The figure shows an edge dislocation with an extra plane of atoms above. This is called a positive (+ 've') edge dislocation. The direction of this extra plane of atoms is the plane of the dislocation. Slip occurs at right angles to this plane along a plane intersecting the dislocation plane at the last extra atom. This is called the slip plane. The dislocation can also move up or down and this movement in the plane of the dislocation is known as the dislocation climb. This is possible by adding or removing an extra row of atoms to or from the dislocation plane.

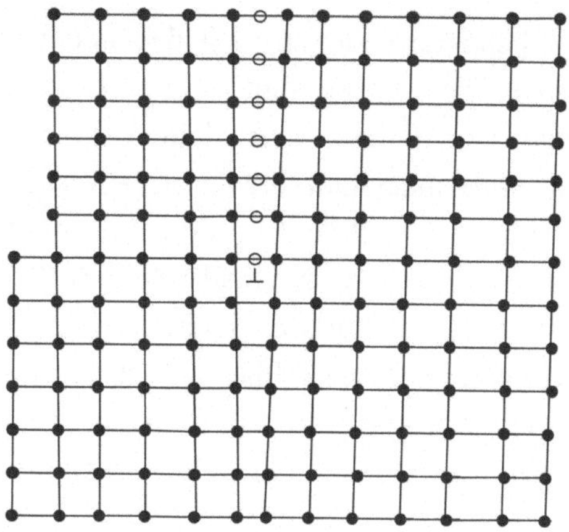

Fig. 4.13 Edge Dislocation.

Another type of dislocation is the screw dislocation (Fig. 4.14). The upper part of the crystal to the right of *AD* has moved relative to the lower part in the direction of the slip. No slip has taken place to the left of *AD*. *AD* is thus the dislocation line. This is parallel to the slip direction and hence this dislocation is called a screw dislocation. Physical appearance of the 'disturbed' region of the atoms also resembles a screw helix and hence the name.

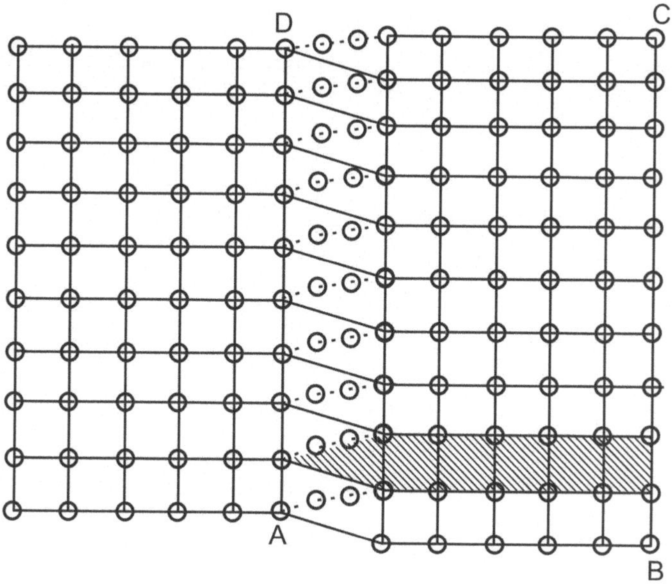

Fig. 4.14 Screw Dislocation.

Metal (or alloy) behavior under stress is dependent upon the effect the stress produces on it. This is the sum of the effects on the individual grains of the metal. So, to have a fundamental concept of the metal deformation, we need to understand the deformation of the single crystal. The slip undergone by the single crystal depends upon the magnitude of the shearing stress produced by the load applied, the geometry of the crystal structure and the orientation of the active slip planes with respect to the shearing stresses. Slip starts when the shearing stress on the slip plane in the slip direction reaches an optimum value called the critical resolved shear stress. This value is really the single crystal equivalent of the yield stress in an ordinary stress-strain curve. The value of the critical resolved shear depends upon the composition and temperature.

Schmid postulated that different tensile loads are necessary to produce slip in single crystals of different orientations can be rationalized by critical resolved shear stress. It is necessary to know from X-ray diffraction, the orientation with respect to the tensile axis of the plane on which slip first occurs and the slip direction of the single crystal tested in tension. Let us consider a cylindrical single crystal of cross sectional area A (Fig. 4.15). The angle between the normal to the slip plane and the tensile axis is 'ϕ' and the angle which the slip direction makes with the tensile axis is 'λ'. The area of the slip plane inclined at an angle 'ϕ' will

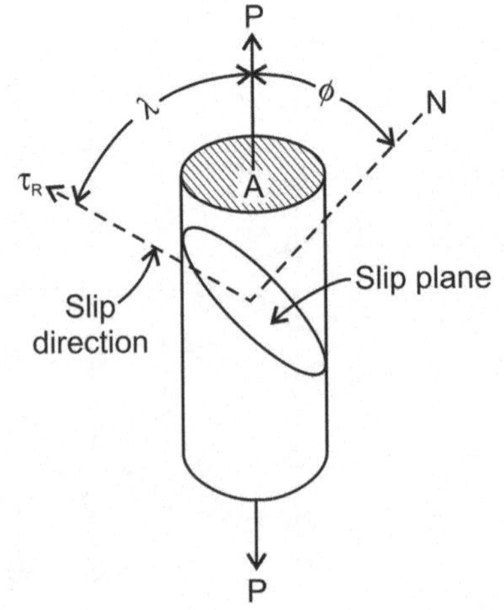

Fig. 4.15 Cylindrical single crystal.

be A/cosϕ and the component of the axial load acting on the slip plane in the slip direction is P cosλ. Therefore, the critical resolved shear stress is given by

$$\tau_R = \frac{P\ cos\ \lambda}{\dfrac{A}{cos\ \phi}} = \frac{P}{A}\cos\ \phi\ cos\ \lambda$$

The above equation gives the shear stress resolved on the slip plane in the slip direction. This shear stress is a maximum when $\phi = \lambda = 45^o$, so that $\tau_R = \frac{1}{2}\left(\frac{P}{A}\right)$. If the tension axis is normal to the slip plane $\lambda = 90^o$ or if it is parallel to the slip plane ($\phi = 90^o$), the resolved shear stress is zero. Slip will not recur for these extreme orientations since there is no shear stress on the slip plane. Crystals close to these orientations tend to fracture rather than slip.

It was experimentally observed that small amounts of impurities in the metal increase the critical resolved shear stress. Addition of alloying elements was found to be of even greater effect. Still greater would be the rise in the value of the critical resolved shear stress in an alloy wherein the solute atoms differ considerably from the solvent atoms in size.

The magnitude of the critical resolved shear stress of a crystal is determined by the interaction of the population of dislocations with each other and with defects such as vacancies, interstitials and impurity atoms. It should be understood that when a positive dislocation moves and encounters a negative dislocation, they cancel out. This stress is of course greater than the stress needed to move a single dislocation yet much lower than the stress required to produce slip in a perfect lattice. On this basis, the critical resolved shear stress should decrease as the density of defects decreases, provided that the total number of dislocations is not zero. When the last dislocation is also eliminated, the critical resolved shear stress should rise abruptly. This rise will be equal to the high value predicted for the shear strength of a perfect crystal.

Most studies of the mechanical properties are made by subjecting the crystal to simple uniaxial tension. In the test, the movement of the cross head of the testing machine constrains the specimen at the grip since the grips must remain in line. As a result, the specimen is not permitted to deform freely by uniform glide on every slip plane along its gauge length (Fig. 4.16). Instead, the slip planes rotate towards the tensile axis since the tensile axis of the specimen remains fixed as in (b). These

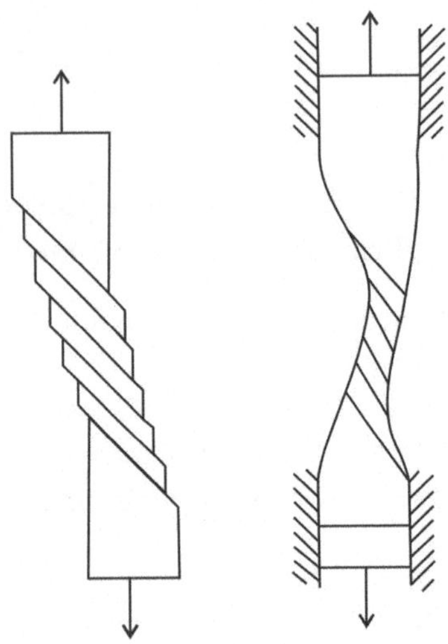

(a) (b)

Fig. 4.16 (a) Tensile deformation of single crystal without constraint
(b) Rotation of slip planes due to constraint.

experimental errors can be compensated by tedious calculations, and it is finally concluded that these can be ignored because of the multitude of grains occurring in the metal specimen under test. This basic concept of deformation at the single crystal level goes a long way in understanding the breakage of the specimen.

FRACTURE OF BRITTLE MATERIALS

Griffith's Theory

Griffith conceived a brittle material to consist of a large number of microcracks. Each of them is capable of producing a stress concentration locally, which equals the σ_{max} of the metal under loads much lower than required ordinarily. If this happens at the region of one microcrack, a crack will form and propagate under favourable conditions. Again, a crack will propagate if the required surface energy for the creation of the two new surfaces (of the crack) 2γ is available. This is obtained by the release of the elastic strain energy hitherto stored in the solid.

Griffith's crack was modelled as an ellipse, the major axis of which is taken as the length of the crack. If the crack is a surface crack, it is taken as half the major axis (Fig. 4.17). The crack propagation stress can be

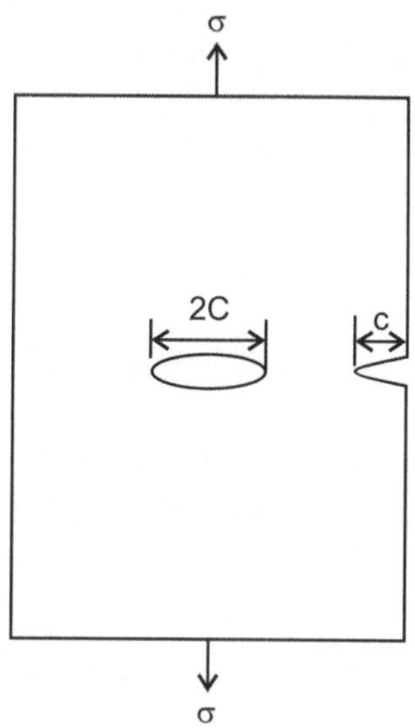

Fig. 4.17 Griffith's crack.

expressed as a function of the size of the microcrack, thus

$$\sigma = \left[\frac{2E\gamma}{\pi c}\right]^{1/2} \quad ...(4.8)$$

where c is half the length of the major axis of the ellipse (half the length of the crack).

We can see that if we increase the crack length four times [c is under the square root in the expression (4.8)] the stress value can be halved as they are inversely related.

Orowan's Modification of Griffith's Expression

It sounds somewhat abstract to imagine a metal fracturing without even a trace of plastic deformation. Perhaps this prompted the various investigators to probe the brittle fracture by fractography (examination of the fracture at high magnifications) and X-ray diffraction. It is established that even a brittle fracture has a plastically deformed region, of course, only a few atoms thick. Hence, Orowan introduced a correction factor. This factor takes into account the plastic deformation, in Griffith's expression. The expression is thus modified as,

$$\sigma = \left[\frac{2E(\gamma + p)}{\pi c}\right]^{1/2} \qquad \text{...(4.9)}$$

where p corresponds to the plastic deformation.

CONCEPT OF CRACK EXTENSION FORCE

We have seen while studying Griffith's fracture theory that the crack will form when an energy equal to the value that is required to create two surfaces (2γ) is available. This is given out, as the thereto-stored elastic strain energy, once the fracture starts. Thus, the reversible elastic strain energy is used up in the irreversible process of crack initiation. This is expressed as the 'elastic strain energy release rate' in terms of in lbs per square inch (or mm kgs per mm^2). This is also called the crack extension force. This is a quantity of importance and is claimed to be a material property independent of the size effects but only governed by composition, microstructure, rate of loading and temperature, just as any other material property.

Mathematical expressions show that this term is analogous to the plastic work factor p in Orowan's modification of the Griffith's eq. (4.9).

CONCEPT OF DISLOCATIONS

It is an established fact that dislocations move under conditions of an external load. The dislocation movement is obstructed by :

(i) the presence of second phase particles,

(ii) the grain boundaries, and

(iii) interlocking by another dislocation.

The obstruction to the movement of dislocations is called a dislocation pile-up. The dislocation pile-up may produce a sessile dislocation, a vacancy, and so on. Detailed studies by eminent workers like Cottrel, Zener and Stroh have established the relationship between grain size, Burger's vector, and the cohesive strength of the metal,

$$\sigma = K.D^{1/2} \quad \text{...(4.10)}$$

where σ = fracture stress,

D = grain size of the metal,

and

$$\sigma = 2\gamma/n.b \quad \text{...(4.11)}$$

where σ = crack propagation stress,

γ = surface energy,

n = number of dislocations interlocked to give rise to a vacancy, and

b = Burger's vector.

With the above study as the background, we can arrive at the following conclusions regarding the fracture behavior of the metals and alloys.

1. The number of dislocations taking part in the dislocation pile-up mechanism leading to the vacancy (crack) formation, controls the metal behavior. If this number (n) is more, the metal behaves as 'brittle'.
2. The grain size of the metal also plays an important role. There appears to be a critical value of the grain size (D) below which the metal fracture will be ductile, and vice versa.
3. There is a temperature range below which the metal fracture will be brittle, and vice versa. This temperature range is called the transition temperature range (correlate to the transition temperature range under impact testing).
4. Higher values of surface energy (γ) tend to promote a ductile fracture. In cases when the surface energy is lowered by the presence of an incongruent precipitate, the metal fracture will be brittle (e.g., hot shortness, H_2-embrittlement).
5. Presence of a notch (c) greatly increases the tendency for a brittle fracture (notch-sensitivity).

References

A.S.T.M., *Special Technical Publication*, 463.

A photographic study of the origin and development of fatigue fractures in Aircraft structures, Ministry of Aviation, U.K., 1961.

Cottrell, A.H., *Trans. Met. Society* AIME, Vol. 212, pp. 192-203, 1958.

Petch, N.J., The fracture of metals, *Progress in Metal Physics*, Vol. 5, Pergamon Press Limited, London, 1954.

Orowan, E., Fatigue and fracture of metals, in the *Symposium at Massachusetts Institute of Technology*, John Wiley and Sons, New York, 1950.

Zenor, C., *Micro-mechanism of fracture in fracturing of metals*, A.S.M., Metals Park, Ohio, 1948.

CHAPTER 5

Fatigue

Like humans, metals too suffer from fatigue. Being humans, we take rest and some food when tired. Metals and alloys do not have that facility and consequently succumb to the fatigue. That is how we can give a concept of the fatigue phenomenon, from a layman's viewpoint.

Fatigue of metals is the property by which they fail at a relatively low value of stress when the stress is repeated. Just as we have seen in the case of impact where the material breaks at a low value of stress (compared to the tensile strength of the material), fatigue failure also occurs at a low stress value. But fatigue is a time-dependent process. An engineering component which is worked continuously for some months or years, whatever the case may be, gives way without any alarm and all of a sudden.

Though the phenomenon is known for over a century, it attained due importance only after the World War II. The value of a fatigue failure can best be appreciated if the reader imagines himself in the position of a lone driver on a highway and the mainshaft of his car is fractured. In the case of an automobile it is only a matter of inconvenience whereas in air and sea travel it is a question of hundreds of lives.

Repeated Loadings

All the engineering components are subjected to loads during service. These loads need not necessarily be steady. In a majority of cases, their magnitudes vary. The material fails, once these fluctuating stresses reach sufficiently high values. The value of this stress may not be anywhere near the static strength of the material. This limiting stress which causes fatigue failure is called endurance limit.

The variation in the stress is generally expressed in two ways :

1. By stating the range of stress that is to be experienced by the metal. This indicates the maximum and minimum values of the stress that the metal will experience. The ratio between the two is known as the 'range ratio'.
2. When the stress is fluctuating, a mean value of the stress should be specified along with the alternating stress to be superimposed to produce the given variation in the stress condition. Thus, the maximum values of the stress can be calculated by superimposing the stress range over the mean value.

A classification of the types of the repeated stresses is given in Fig. 5.1. This classification is based on the report of the ASTM research committee on fatigue of metals. In all these classifications, however, the nature of the stress, viz., tension, compression, torsion, etc., should be specified.

A common type of engineering design, in which each of its components should be designed taking into account its fatigue strength, is a road or rail bridge. When a lorry or the rail is moving on the bridge, the structural members are subjected to repetition of the loads. Further, at any single point on a railway bridge, the loads become cyclic between the passage of the wheels across the point. In many other cases, stresses are quite severe and more frequently repeated. The main shaft and other part of a steam or oil engine, springs, wire ropes, crane hooks, car-axles, rails, railway axles, etc., come under this category. Moreover, the loading experienced by them is mostly cyclic.

Fatigue Strength and Endurance Limit

In a fatigue test, the value of stress at which the metal fails is called the fatigue strength. But in practice, the components are never designed to develop that value of stress which is equal to the fatigue strength by employing factors of safety in design. A limiting stress calculated from the endurance limit is always taken into account. The endurance limit is determined for a particular number of cycles of stress.

Fatigue Test Loading Conditions

Of the types of repeated stresses described in Fig. 5.1, the totally reversed stress is the most severe one. That is why fatigue strength of metals is generally determined under completely reversed bending conditions. Thus, the values or data are always taken to be that of completely reversed bending test only, unless otherwise specified.

TYPES OF REPEATED STRESSES*

TYPES OF STRESS VARIATION		RANGE RATIO NOMENCLATURE		MEAN STRESS NOMENCLATURE	
DESCRIPTION	DIAGRAM	MAX. STRESS	RANGE RATIO	MAX. STRESS	ALTERNATING STRESS
STEADY STRESS σ_1	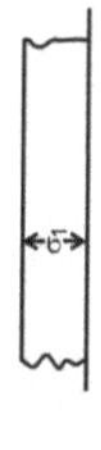	σ_1	$\frac{\sigma_1}{\sigma_1} = 1.0$	σ_1	O
PULSATING STRESS BETWEEN σ_1 AND σ_2		σ_1	$O < \frac{\sigma_1}{\sigma_2} < 1$	σ_m	$\pm\sigma_a$
PULSATING STRESS BETWEEN σ_1 AND O		σ_1	$\frac{O}{\sigma_1} = O$	σ_m	$\pm\sigma_a$
PARTLY REVERSED BETWEEN σ_1 AND (−) σ_2 WHERE $\sigma_2 < \sigma_1$ AND OF OPPOSITE SIGN	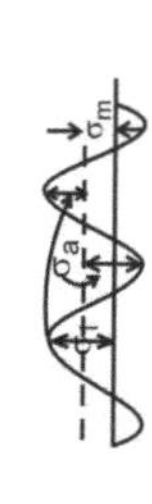	σ_1	$-1 < \frac{-\sigma_2}{\sigma_1} < O$	σ_m	$\pm\sigma_a$
COMPLETELY REVERSED STRESS BETWEEN σ_1 & σ_2 WHERE $\sigma_1 = \sigma_2$	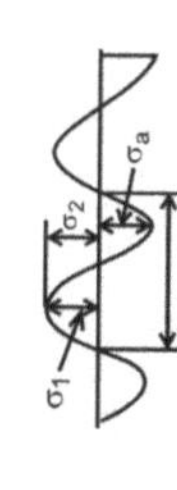	σ_1	$\frac{-\sigma_2}{\sigma_1} = -1.0$	O	$\pm\sigma_a = \sigma_1$

NOTE : $\sigma_m = (\sigma_1 + \sigma_2)/2$ & $\sigma_a = (\sigma_1 - \sigma_2)/2$

**From Proceedings A.S.T.M., Vol. 37, Pt. I, 1937.*

Fig. 5.1 Types of repeated stresses.

In general design, the endurance limit is taken as a fraction of the tensile strength. For steels this value lies somewhere between 0.4 to 0.6. It is lower than 0.4 in the case of non-ferrous metals and alloys. The supported beam. In the cantilever type again, the load may be applied at a single point or halved and applied at two places, constituting a couple. When the load is applied at a single point, there is a tendency for the bending moment to concentrate near the fixed end or chuck. This is minimized by giving a suitable taper to the specimen. On the other hand, when the loading is done at two places, a larger volume of the test specimen is brought under the influence of the load (see Figs. 5.2 and 5.3).

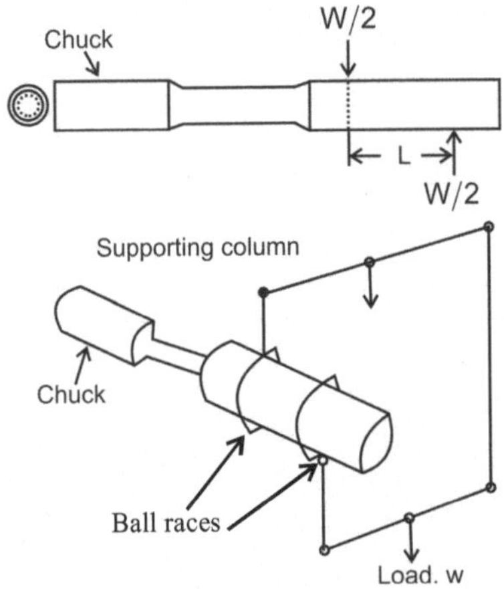

Fig. 5.2 Two-point loading rotating beam fatigue testing.

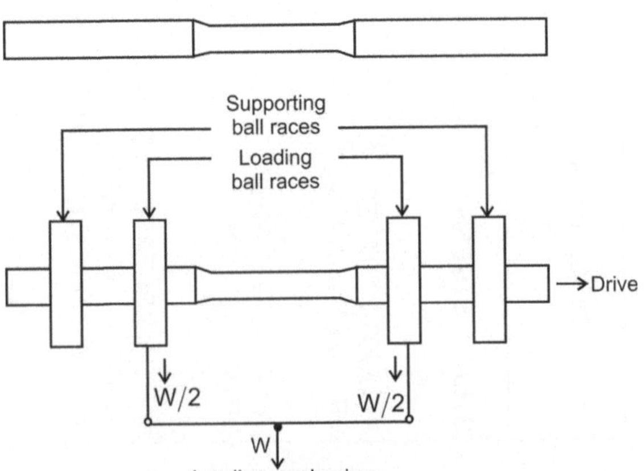

Fig. 5.3 Four-point loading rotating beam fatigue testing.

The simply supported beam type loading is made use of in the Simplex rotating beam machine. The test specimen has an effective diameter of 7.62 mm (0.3″). The long shoulders of the specimen are for mounting it on two sets of ball races, one on each side. The outer races serve as the supports and the inner ones load the specimen [see the line diagram (Fig. 5.3)]. The specimen is rotated by an electric motor. The Baldwin Sonntag rotating bend fatigue testing machine also operates on the same principle but it has a higher capacity. It can test specimens up to 1″ (25 mm) effective diameter.

A large variety of fatigue testing machines are in the market. A thorough review of the various machines is given by Forrest.[2] Special machines are developed by national testing laboratories or such bodies or large industries in many countries for special and specific fatigue testing purposes of paramount importance. The National Engineering Laboratory and the London Transport Board (both U.K.) have such huge machines for bend fatigue testing of specimens as big as railway axles (Fig. 5.4).

Fig. 5.4 Heavy duty fatigue equipment (specially designed for a research project).

(*Courtesy* : London Transport Board, Chiswick, U.K.)

Similarly the Moscow Central Scientific Research Institute of the U.S.S.R., Ministry of Transport, have machines capable of testing axles up to 12″ (300 mm approx.) diameter.

Summarizing, the various types of fatigue testing machines incorporate the following:

1. A means of holding the specimen suitably.
2. A means of loading and measuring the load.
3. Provision to rotate the specimen and measure the number of revolutions.*
4. A means of disconnecting the revolution measuring device or putting off the machine itself, once the specimen is broken.

Machine designs differ in respect of the first two features. The specimen may either be loaded as a beam or cantilever. Again, the loading may be simply by mounting weights onto a weight holder or by hydraulic means or attracting and leaving a temporary magnetic head by means of a strong and soft electromagnet (The Amsler's Vibrophore) or by means of an eccentric or crank.

A fatigue testing machine manufactured by M/s. Avery Denison Limited, Leeds, U.K., is shown in Fig. 5.5. The grips for use on combined bending and torsion actions are separately shown in Fig. 5.6. The standard test specimen for use on the machine is shown in Fig. 5.7.

**Note :* A recent feather in the cap of fatigue researchers is the work of the British Non-ferrous Metals Research Association. They developed a new design for bend fatigue testing of metals at elevated temperatures. In this, the specimen remains stationary and the load is rotated. The stationary specimen facilitates an easy access for temperature measurement.

Fig. 5.5 Laboratory type fatigue testing machine.
(*Courtesy* : Avery-Denison Ltd., Leeds, U.K.)

Combined bending and torsion

Bending

Fig. 5.6 Grips for bending of round specimens and grips for combined bending and torsion of round specimens.
(*Courtesy* : Avery-Denison Ltd., Leeds, U.K.)

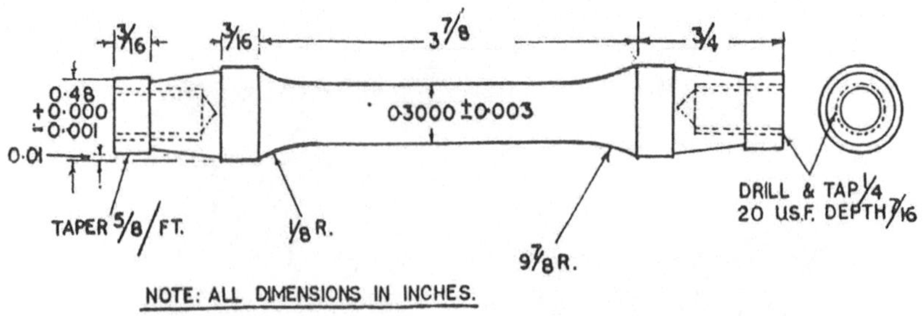

Fig. 5.7 Standard specimen for rotating beam test.
(From *Metals Handbook*, A.S.M., 1948).

TEST PROCEDURE

A number of standard test specimens are made from the metal under test. The first specimen is tested at a high value of load. The number of revolutions the specimen experiences before fracturing is noted on the counter. Another specimen is fixed in the machine and this time, the load is slightly decreased. The number of revolutions that are indicated in the counter this time, will be more than that in the previous case. The other specimens are also broken one after the other, each time reducing the load, and noting the number of revolutions. The procedure is continued until a value of stress is reached when the specimen does not fail, say even after 10 million revolutions.

When the specimen is rigidly fixed at one end and loaded at the other, it bends. In such conditions, the top half of the layers in the specimen experience a sort of a stretching action while those in the bottom half experience compression. In Fig. 5.8 (b), the layers experiencing stretching or tension stress are white while those experiencing compression are shaded. Once the specimen turns half a revolution, it literally comes into an upside down position with respect to its initial condition. In such a condition the stress distribution in the specimen also changes exactly in an opposite fashion. That is, those regions which were under tension become compressed and vice versa [Fig. 5.8 (c)]. Further rotation through 360° brings the specimen back to its initial position. Thus, during one complete revolution, the stresses in the specimen are completely reversed once. It means that, by measuring the number of revolutions the specimen is subjected to before failure, we actually measure the number of reversals of the load.

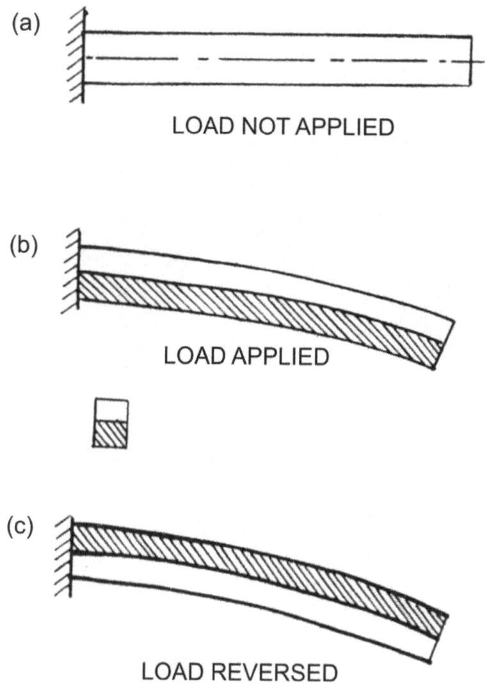

Fig. 5.8 Stress distribution in a cantilever loaded and rotated.

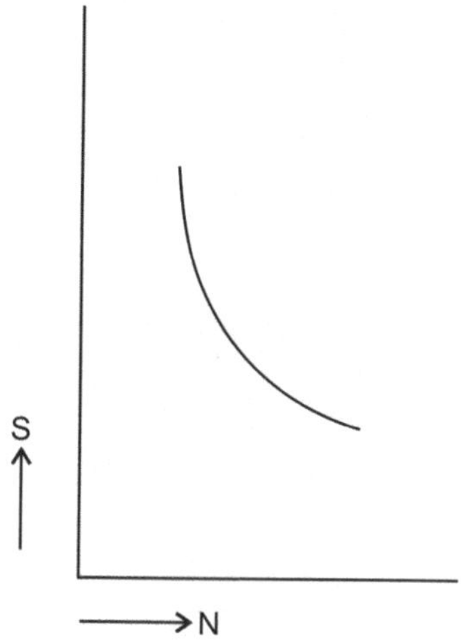

Fig. 5.9. S-N diagram.

From the data thus obtained, the stress, number of reversals of the stress diagram, commonly referred to as the S-N diagram, is plotted (Fig. 5.9). The stress value is plotted on the Y-axis and the number of reversals of the load, taken on a logarithmic scale, on the X-axis. The stress value at which the curve becomes nearly horizontal, is taken to be the endurance limit of the metal.

THE FATIGUE FAILURE FRACTURE

A typical fatigue fracture surface is shown in Fig. 5.10. The fracture is characteristic and easily recognizable because of the following:

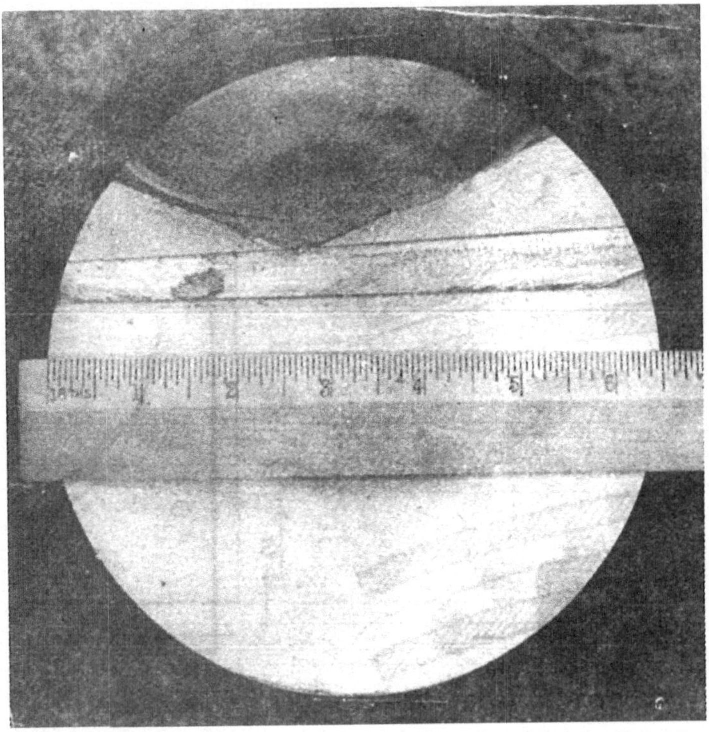

Fig. 5.10 Photograph illustrating a typical fatigue fracture in a large specimen.
(*Courtesy* : London Transport Board, Chiswick, U.K.)

1. Roughly speaking the fracture occurs suddenly, with no indication or warning.
2. Visually, the direction of fracture is at right angles to the direction of the tensile stress.
3. Very often, the fracture shows, up to a certain depth, a series of rings originating from the point of initiation of the failure. These rings are called the 'beach marks'. At this region, the fracture resembles the concoidal fracture generally observed in glass and minerals.

4. The fracture transforms into a ductile type in regions beyond those mentioned in (3). This shows that the part was unable to withstand the load once the crack has propagated to a considerable depth.
5. Above all, the point of origin of the fracture is clearly revealed. Generally the fracture starts at a surface defect or irregularity like a keyway, sharp corner, notch, root of a screw thread or an inclusion, which causes a stress concentration (Fig. 5.11).

Fig. 5.11 Fatigue cracks in a large machine gear. Note that the crack has initiated at the root of the gear.
(*Courtesy* : Magnaflux Corporation, Chicago, Illinois, U.S.A.)

Theories

There are many theories explaining the fatigue phenomenon. But not one of them is totally comprehensive and conclusive. However, failure due to the building up of the 'slip lines' at a particular region appears to be a reasonable proposition.

Orowan's Theory

According to Orowan, metals consist of minute 'weak' regions. The deformation in these regions is more than in the neighbouring 'strong' regions, under repeated loading. The strain increases as the value of the load increases. When the load or stress applied is such that it exceeds the value which the weak region can sustain, a crack develops.

Orowan's 'weak' regions may be those regions where the orientation of the grains is favourable for slip. They may also be the areas of high stress concentration like cracks, roots of the notches, screw threads.

Again, when a crack is developed it becomes another weak region. The process repeats itself resulting in the formation of fresh cracks. These get aligned suitably and result in a larger crack, leading to fracture.

Wood's Concept

When loaded, metals deform under conditions of slipping (Fig. 5.12a). When the load is reversed, the direction of slipping of the metal is also reversed (Fig. 5.12b). As a result there is every possibility for a notch to develop when the load reversal is repeated many times. This notch may be the root cause for the final fatigue failure.

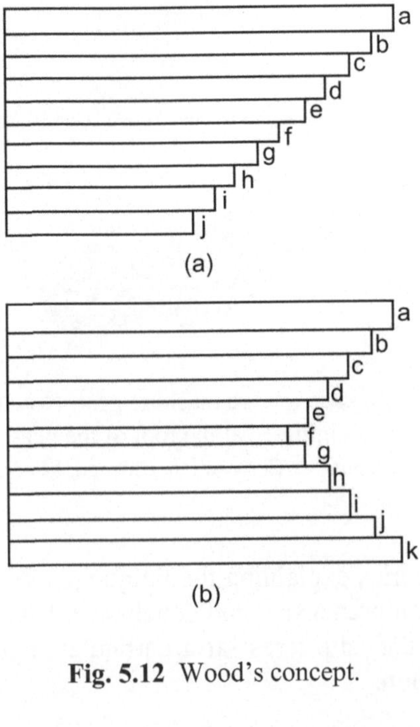

Fig. 5.12 Wood's concept.

This theory sounds reasonable because it establishes that – (1) the fatigue failure starts from the surface, and (2) the cracks initiate at the slip-bands, intrusions and extrusions.

EFFECT OF IMPORTANT VARIABLES

The following factors have a pronounced influence upon the fatigue property of metals:

1. Composition,
2. Stress concentration,
3. Size,
4. Surface condition and
5. Mechanical working.

1. Composition

The fatigue strength of a metal or alloy is not directly influenced by the alloying elements. However, those which increase the static (tensile) strength are found to increase the fatigue strength also. In the case of steels, carbon is found to have the maximum effect on the fatigue strength; carbon increases the fatigue strength.

It is further established that the alloys which undergo strain-ageing (work-hardening) exhibit a sharp knee in the S-N curve. In the case of steels, the sharpness of the knee is found to decrease and the curve flattens out as the carbon and nitrogen contents are decreased. Absence or decrease in the two alloying elements which are interstitial in character, means that the steel is less prone to work-hardening.

2. Stress Concentration

Fatigue strength is seriously affected by the presence of stress raisers in the specimen. Notches, holes, keyways, etc., which are found to initiate the fatigue failure, are all stress raisers. But all the engineering components invariably contain these stress-raisers. It is impossible to manufacture a bolt without threads, and without the threading, the bolt itself is useless. However, it is advisable to avoid those that are avoidable.

Besides, the changes in section and surface irregularities like machining cracks, porosity, inclusions, decarburised regions, etc., have their own influences in this regard.

A surface irregularity will act similar to a notch. The effect of a notch on the specimen is discussed in detail under impact testing. The stress concentration is increased at the root of the notch and thus a gradient of stresses is set up from that region towards the axis (centre) of the specimen.

The effect of a notch on the fatigue strength can be found out by testing sets of the similar specimens of the same material, one set notched and the other unnotched. The ratio between the endurance limit of the unnotched specimens and that of the notched specimens can be computed to give what is known as the 'fatigue notch factor K_f'. It is experimentally established that this factor varies with:

(i) the severity of the notch,
(ii) the shape of the notch,
(iii) the stress value,
(iv) the type of the stress (or loading), and
(v) the material.

3. Size

Unlike tensile strength, the fatigue strength is dependent upon a great many factors. Most important of them is the reversal of the stress. Though in all stress reversals the stress cycle changes from a tension peak to the compression peak, its effect is more severe on a thin section than in the case of a thick object. In the illustration (Fig. 5.13), value of *d* (the distance between surface and the axis) is less than *D*. Thus, the stress (or strain) gradient from the surface to the centre in the case of a thicker specimen is more. Also the volume of the material deformed is more. Further, increased size results in an increased surface area. Thus, there is more likelihood that a thicker section will fail.

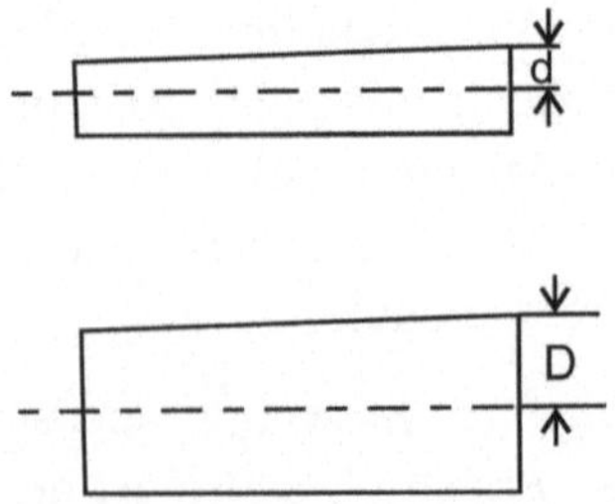

Fig. 5.13 Specimens of different sizes.

An exacting study of the effect of size on the fatigue is very difficult, for, it is practically impossible to prepare geometrically similar specimens of varying sizes having the same metallurgical characteristics like the structure and stress distribution. Loading thicker sections is another difficulty, which demands machines of greater capacities. However, the effects are studied within the limitations and it is established that the thicker the section, the lower is the fatigue strength. The fatigue strength is found to decrease by 25% when the thickness of the specimen is increased by 1/4″ above 1/2″. However, no marked decrease was noticed above the section size of 3/4″. It is further seen that in the case of steels, the size effect became more prominent as the carbon content is increased.

4. Nature of the Surface

Surface is the source wherefrom the fatigue failure originates. The condition of the surface has a pronounced effect on the fatigue strength of a metal.

It is established that different surface finishes produce different fatigue strengths for the same metal. The smoother the surface is, less is the scope for the presence of machining cracks or other surface irregularities. The surface roughness* is expressed in microns. As the surface roughness figure decreases the fatigue strength increases. In general, a buffed surface can be regarded as smooth, and an ordinary machined surface as rough. The fatigue strength of the same steel is observed to be 20-25 % less when tested in the ordinary machined condition than when the specimen is finished by buffing.

It is observed that the fatigue strength of a metal or alloy is remarkably increased if its surface is made harder. Logically it stands to reason, for it is difficult for a hard region to give way, and hence the scope for the initiation of the first fissure is minimized. Steels are case-hardened for improving their fatigue strength.

In general, electroplating is found to decrease the fatigue strength of steel. However, soft cadmium platings are made with no detriment to the fatigue strength. The effect is governed by many other factors in the platings, such as the nature of the plating, adhesion, residual stresses, induced stresses, etc.

---

* Separately treated under surface measurements.

5. Residual Stresses

Residual stresses are those stresses which still remain in the object after the process of manufacture is completed. These have very low magnitudes but at the same time manifest themselves very dangerously, for, they act on microscopic areas.

The origin of the residual stresses may be:

1. plastic deformation during which the material flow is not uniform throughout,
2. improper and insufficient annealing or normalizing after the plastic deformation,
3. quenching,
4. improper and insufficient tempering after quenching,
5. some finishing operations like machining, threading, boring, etc., and
6. sudden heating and cooling or excessive momentary loading, leading to improper recovery during the service.

It is said that the nature of the residual stress in a plastically deformed metal will be opposite to that by which the deformation was made. A metal which is plastically deformed in tension will consist of residual stresses of compressional nature and vice versa. For all practical purposes, the residual stress can be considered as identical to an externally applied stress. The maximum intensity of the residual stress in a metal is equal to its elastic limit.

A compressive residual stress pattern at the surface of an object increases its fatigue strength remarkably. This is the most effective method by which the fatigue strength of a metal can be improved. This is accomplished by:

(i) shot peening,

(ii) surface rolling using contoured rolls, and

(iii) in the case of a single object, the surface is just hammered lightly by a hand hammer.

It goes without saying that excessive peening or rolling will damage the fatigue strength instead of improving it. A series of laboratory tests and experience are warranted before a process is standardized for routine production work.

Another point that should be noted in this connection is that not all the aforesaid processes yield residual stresses of compressive nature. Those processes which give rise to tensile stresses should be avoided. If imminent, it should be followed by another treatment to improve the fatigue strength. Grinding can produce both the tensile and compressive types of stresses. Any process should be carefully controlled to avoid tensile type residual stresses.

INTERPRETATION OF FATIGUE TEST DATA

Fatigue test results obtained from the laboratory testing of polished small fatigue specimens cannot be taken as the sole basis for engineering design. It is opined that the fatigue test should be made on sample 'parts' preferably than on standard specimens. Further, it is suggested that the testing be performed, if possible, under the stress cycles similar to those the parts are expected to experience during their life. The number of stress cycles the part is tested for is arrived at by a prior estimate of the service life of the engineering component.

Again the test results should be considered only as a basis for design and not as statements of precision. Over and above the test data, comes the factor of safety. The factor of safety is fixed to take over the uncertainties like the corrosive media and the stresses the engineering member is likely to withstand during its life.

In laboratory tests on actual structurals and machine parts, loads of various magnitudes can be employed so that the range of stress during a stress cycle will approximate the stresses experienced by the part in actual usage. Thus, different special purpose fatigue testing machines are designed, differing in designs and capacities depending upon the purpose.

Tests of the aforesaid nature are both cumbersome and costly. That is why attention is again focused on the laboratory test data. Conventional fatigue test data can be relied upon for design purposes by suitably adapting it for the particular purpose. This adaptation includes making allowances for size, shape, surface condition, presence of stress raisers, corrosive atmosphere and the stress range in the actual service.

For the purpose of arriving at the due allowances, the criterion which is called the 'strength reduction factor' is determined. Reduction of the strength due to the variations in the shape or other things is thus duly taken care of. A series of fatigue tests are made on both notched and unnotched test specimens keeping the critical section the same as that of the engineering member. *S-N* curves are drawn for these tests. The ratio of the endurance limit for the unnotched specimen to that of the notched specimen is calculated, which gives the strength-reduction ratio. This is true for the particular size and shape of the notch and the given critical size of critical section.

Another design consideration is the 'stress concentration factor', a discussion about which falls out of the track in the present discussion. However, in general, it is found that the strength-reduction ratio is less than the theoretical stress concentration factor and so the former can be safely followed in design.

With all the above, the effect of size and surface condition should always be taken into account and due consideration shown for them.

Fatigue Failure in Service

Fatigue failures are most common in shafts – turbine shafts, main shafts of automobile and naval vessels, etc. The keyways, changes in section, screw threads and oil-holes are the general originators of the cracks and the subsequent fatigue failure. Very interesting fatigue failures occurred in the mid-fifties due to which two B.O.A.C. Comet aircraft were lost. The plane wreckages recovered from the sea beds were investigated into and it was determined that the wreckage resulted due to the fatigue failure of the fuselage. The stresses responsible were assumed to be those induced by the cabin pressurizing cycles for each flight. It was assumed that the cracks originated at the corners of the aerial windows.

Corrosion Fatigue

The fatigue failure, occurring in a corrosive atmosphere, is referred to as 'corrosion fatigue'. In a corrosive atmosphere the fatigue failure is hastened. Without any stress application itself, let apart the stress cycles and reversions, the metal develops surface fissures or pits in a corrosive medium. When corrosion and fatigue are simultaneous, the metal fails in a very short time and at a low value of the stress.

Many methods are recommended for improving the corrosion fatigue of metals and alloys. All of them aim at increasing the corrosion resistance rather than the fatigue strength. While selecting materials for service under corrosive conditions, stainless steels, bronzes, copper-beryllium alloys, etc., which are highly corrosion-resistant are recommended. In cases where such a choice is not possible, i.e., when steels only are to be used, the metal may be protected from the corrosive medium by surface coatings or cladding. Zinc and cadmium coatings and aluminium cladding are commonly employed. At the same time, the nature of the coating should not adversely affect the fatigue property. Other methods recommended are nitriding and shot peening.

In all these cases, it is of paramount importance to bestow greater attention from the design point of view of the object. Sharp changes in section, sharp corners, etc., should be avoided as far as possible. These things decrease the fatigue strength even as they are and will be more harmful in a corrosive environment.

Torsion Fatigue

The advent of torsion bar suspension of automobiles and the requirement of the service condition testing of springs brought out the necessity of torsion fatigue testing. Unfortunately the number of torsion fatigue testing machines in the market is very limited. The Schenck Torsator is one that operates on the principle of mechanical resonance (Fig. 5.14). Torsion fatigue can also be determined by the combined stress fatigue testing machine. This machine, practically a universal fatigue testing machine, can test the fatigue strength under any type of loading by a suitable variation of the specimen mounting.

Fig. 5.14 Vertical resonance pulsator.

(*Courtesy* : Carl Schenck Maschinenfabrick GmbH, Darmstadt, W. Germany.)

METALLURGICAL FACTORS IN REGARD TO FATIGUE

Like tensile strength to which it is very closely correlated, fatigue strength is a structure-sensitive property. Unless otherwise mentioned, the fatigue strength for steels can be taken to be half the tensile strength. However, this fatigue ratio (the ratio of the fatigue strength and the tensile strength) varies for different metals and for the same metal depending upon many factors.

Like tensile strength, the fatigue strength of a metal increases as the grain size decreases. Similarly, solid solution alloying also increases the fatigue strength. In the case of hardened steels, the fatigue decreases as the percentage of martensite formed decreases. Steels are found to possess the optimum fatigue properties in the quenched and tempered condition. Even a trace of surface decarburisation is found to adversely effect the fatigue strength. Thus, fatigue strength is more structure-sensitive than the tensile strength.

In the case of forgings, directionality is exhibited. The fatigue strength is about 20-30% lower if loading is at right angles to the direction of flow.

Temperature and Fatigue

(a) **Below the room temperature.** Tests on steels have shown that the fatigue strength increases with a fall in the temperature of testing below the room temperature. Unlike the impact properties, there is no sudden embrittlement from the fatigue point of view.

(b) **At high temperatures.** Fatigue strength decreases at temperatures above the room temperature. An exception is mild steel. Mild steel shows a maximum fatigue strength in the temperature range of 200°-300°C.

Materials which show a sharp fatigue limit when tested at the room temperature will not record the same at higher temperatures, especially above 450°C. At higher temperatures the effect of creep dominates the fatigue. Creep is discussed in the following chapter.

Thermal Fatigue

Metals and alloys expand on heating and contract on cooling. The amounts by which the expansion and contraction occur are characterized by the nature of the metal; for some it is high and for others it is low.

Expansion means elongation in all directions and its effect can be understood in a way similar to the elongation occurring under a loaded condition. The contraction may not be complete in many cases, especially when the cooling is sudden. This induces residual stresses.

Failure of metals on thermal cycles is referred to as thermal fatigue. In general it can be said that a higher thermal conductivity decreases the possibility of thermal fatigue failure.

Fretting

When two engineering members are fastened to one another in any mechanical fashion such as nut and bolt or rivet, there will be a relative motion between them. During service, they get damaged at the surfaces in contact. This is called fretting.* The surface damage in turn leads to a fatigue failure.

Consequent to the fretting phenomenon, metal is removed from the surface by a grinding action. These particles get oxidized readily at the higher temperature developed owing to friction. A metal oxide powder comes into existence which acts as an abrasive. The presence of such a powder at the contact surface accelerates further grinding of metal surfaces.

The best method to avoid fretting is to minimize, if not eliminate, the relative motion. If the temporary joints cannot be eliminated by welding, a good lubricant should be used. A solid lubricant like molybdenum sulphide whose film will last long on the surface of friction is recommended.

*This is different from wear, since the relative motion of the members in contact will not be at a higher velocity to fall into the class (nature) of wear.

References

Fenner, A.J., *Mechanical Testing of Materials*, George Newnes Ltd., London, 1965.

Forrest, P.G., *The Fatigue of Metals*, Pergamon Press Ltd., London, 1962.

Cazaud, R., *Fatigue of Metals*, Chapman and Hall, 1953.

Sines, George and Waisman, J.L., *Metal Fatigue*, McGraw-Hill, 1959.

Manual of fatigue testing, A.S.T.M. Sp. *Tech. Publication*, 91, 1949.

IS: 5619–1970, Recommendations for Fatigue Testing of Metals.

CHAPTER 6

Creep

Creep is defined as the quasi-viscous flow of the metal. This comparatively occurs at a slow but progressively increasing strain. Creep may be rapid or slow but its rate decreases rapidly as the stress is reduced (this happens at any temperature). But generally the term creep is used to denote the strength of a metal at temperatures much above the room temperature.

High temperature strength of metals is an important requirement. A bearing gets heated up in continuous use, yet it should not soften. The design of moving parts for high speed machinery and many auto and aero engine parts necessitate the employment of materials with good high temperature properties. Rocket technology demands materials with very high and rigid creep requirements. In fact, new alloys have been and are still being developed to meet these challenges of high temperature service requirements.

When a metallic screw cap is jammed we light a match and warm it. The idea behind it is that due to expansion, it gets loosened. Expansion, a property which increases a material's dimensions, will naturally loosen its "joints", i.e., the grain to grain binding strength of the metal is decreased. In other words, the metal becomes plastic or weaker. It is seen in the earlier chapters that the strength of a metal decreases with temperature. This is explained by the greater mobility of the dislocations with the increased temperatures. Further, the atmosphere in which the metal functions, also plays an important part in the creep phenomenon.

Added to the above, the precipitation of hard phases with temperature (ageing) and the grain growth, do exercise their own effects on the metal properties. Thus, the metal properties at the higher temperatures are governed by and depend upon a series of complex variables.

While considering the creep phenomenon from the high temperature serviceability of the metal, strain is as important as the time of exposure. Many metals behave as viscous materials and undergo a time-dependent deformation.

THE CREEP CURVE

A constant stress (load) is applied to a tensile specimen held at a constant temperature. The strain (extension) produced is measured as a function of time and these values are plotted. An ideal creep curve is illustrated in Fig. 6.1.

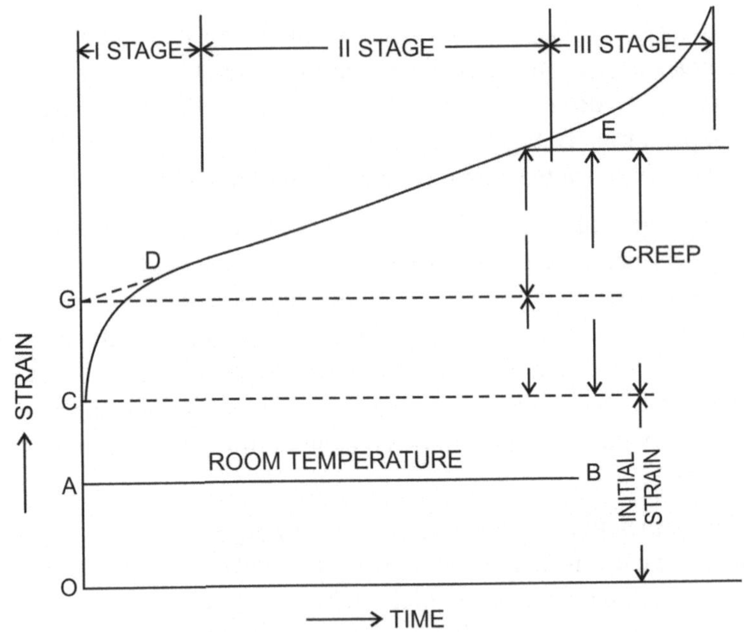

v_0 = Creep rate during II stage
t = Time during II stage
ϵ_0 = Elementary creep
$\epsilon_p = \epsilon_0 + vt$ = Total creep in time t

Fig. 6.1 The creep curve.

The slope of this curve is called the 'creep rate'. When the load is applied to the specimen, there occurs an almost instantaneous elongation. This is denoted by ε_o. The creep rate decreases with time during this period and is known as the primary creep. The creep rate later becomes approximately constant in the second stage. The third stage is marked by a rapid increase in the creep rate, which continues until the fracture occurs.

In actual experiment, the third stage is not attained. We have machines with devices which compensate for the necking phenomenon and thereby maintain a constant stress on the specimens throughout the test Thus a creep curve which appears as an extended second stage curve will be obtained.

Andrade's Concept

Andrade considers that the constant-stress creep curve consists actually of two curves representing two separate creep processes superimposed. The first component is the transient creep which has a decreasing creep rate with time (Fig. 6.2a). The other one is a constant rate creep curve (Fig. 6.2b). When these two are superimposed, the so-called creep curve is obtained (Fig. 6.2c).

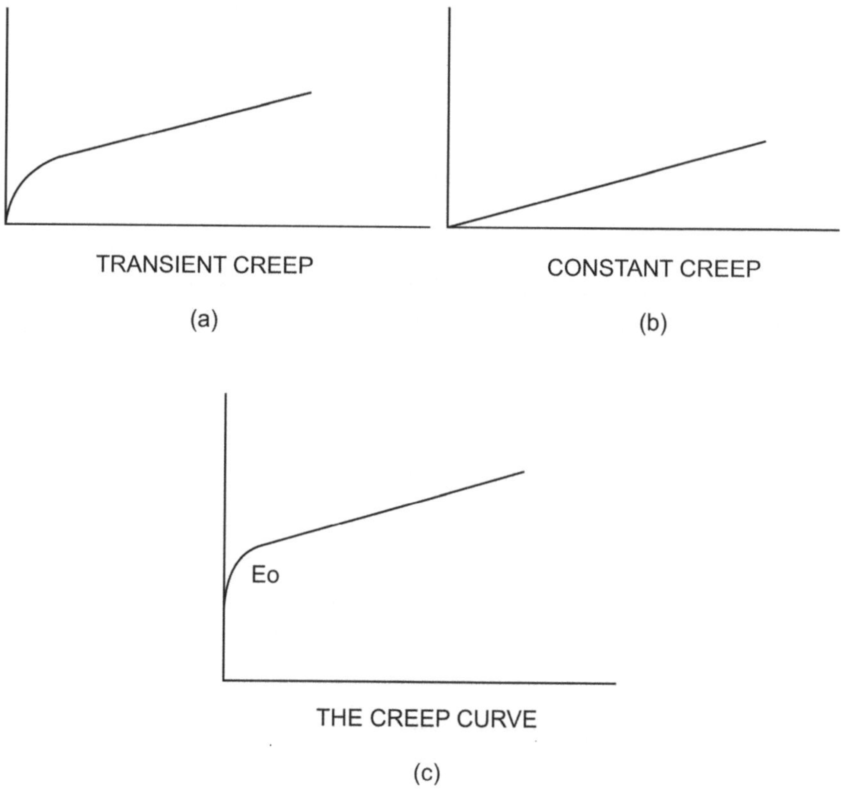

Fig. 6.2 Andrade's concept.

According to Andrade, the creep curve is represented by the empirical equation –

$$L = L_o (1 + \beta t^{1/3}) \exp Kt$$

where L = length of the specimen at time 't', and

L_o, β and K = constants whose values are empirically determined.

L_o approximates the length of the specimen when the sudden strain produced on the application of the load has ceased. The transient creep is represented by β, a constant. Thus, when $K = 0$ and the above equation becomes

$$L = L_o (1 + \beta t^{1/3})$$

$$\varepsilon = dl / dt = (1/3) L_o \beta t^{-2/3}$$

Again, when $\beta = 0$,

$$L/L_o = \exp Kt$$

$$dL / dt = K L_o \exp Kt = KL$$

or, $$K = (I/L) (dL/dt)$$

K thus represents an extension per unit length which continues at a constant rate. It is the viscous component of creep.

The transient creep is often referred to as the β-flow and the steady state creep as the K-flow.

In the primary creep stage, mostly the transient creep predominates. A certain deformation of the material occurs and the material gets work-hardened. Thus, its resistance to flow increases. In the second stage creep process, the rate of deformation and work-hardening and the subsequent resistance to flow are nearly equal. A balance will be set up between the opposing processes of strain hardening and recovery. The average creep rate during this stage is termed as the minimum creep rate. The third or tertiary creep state is predominantly the final breaking stage. The reasons for the increase in the creep rate are not yet known but it is supposed that it is due to the necking phenomenon.

EQUI-COHESIVE TEMPERATURE

At room temperature, metals fail by fracture through the grains themselves. We know that the grain boundaries are the regions of high stress concentration and possess no fixed atomic arrangement. Consequently, they are hard and strong. However, this does not hold good in the case of failure at higher temperatures. The grain boundary region undergoes a remarkable degree of 'mobility' leading to a condition of a thin, weak region. The metal in such a state behaves as a material consisting of strong grains embedded in a thin, weak matrix. This is conducive to the grain boundary failure which precisely occurs in creep. It is termed as the intergranular or intercrystalline failure. There exists a temperature for every metal below which the failure will be through the grains, viz., the fracture is transcrystalline, and above which it will be through the grain boundaries, i.e., intercrystalline. This temperature is known as the "equicohesive temperature". At this temperature, the grains and grain boundaries are equally strong.

Equi-cohesive temperature should not be confused with the recrystallization temperature or the transition temperature or even the transformation temperatures of the steels. However, it is nearly equal to the recrystallization temperature for most of the metals. (The equi-cohesive temperature is about 400 °C for plain carbon steels and slightly higher for alloy steels).

GRAIN BOUNDARY FAILURE

Under the conditions of creep, the grain boundary failure may be due to any of two phenomena (Fig. 6.3) :

(i) grain boundary sliding, and

(ii) void formation in the region.

It is seen above that the grain boundary region is weaker than the grain at the higher temperatures. When a load is applied, a sliding action occurs in the region. In such a condition, a crack develops at the regions where three grains meet. This mechanism of crack formation is prevalent when a high temperature and small loading are employed. On the other hand, if the loads are higher and the temperatures lower, voids form at the grain boundary regions which are at right angles to the direction of stress. Gradually these voids grow and coalesce into grain boundary cracks. Many research workers have described the formation of voids phenomenon, the most interesting of which claims that voids form as a result of grain boundary sliding. Investigations into this mechanism are still going on and nothing definite can be put forward, at present.

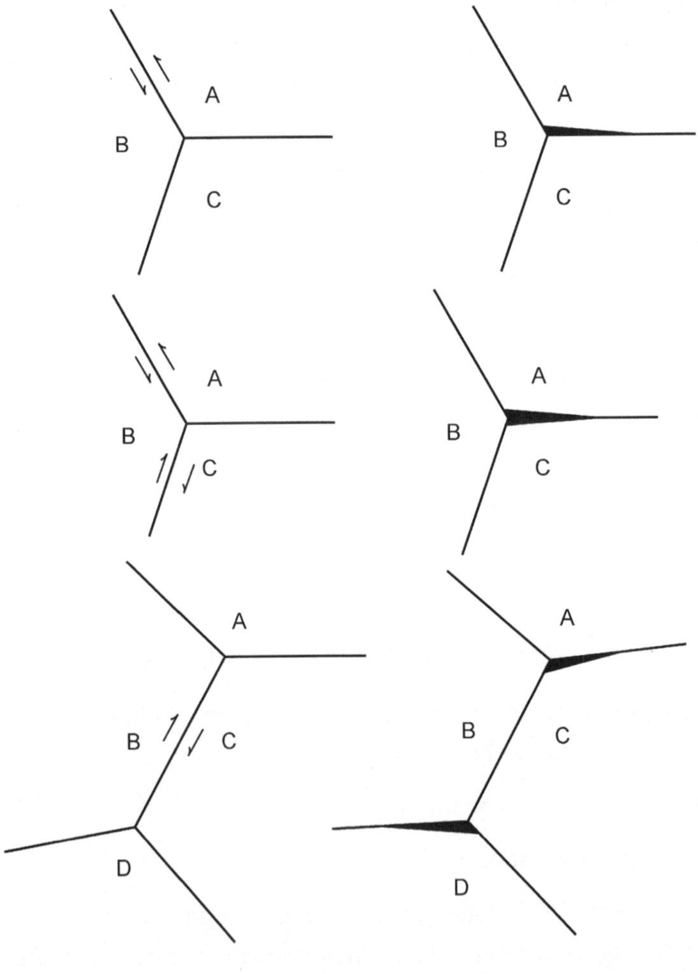

Fig. 6.3 Grain boundary sliding and inter-crystalline crack formation.

(H.C. Chang and N.J. Grant, Trans A.I.M.E., Vol. 206, p. 545, 1956.)

CREEP TESTS

Determination of the creep properties of metals is quite simple. The equipment needed is similar to the one needed for a tensile test. However, the specimen is enclosed in a furnace whose temperature is very accurately controlled and determined. The set-up for creep testing is schematically illustrated in Fig. 6.4.

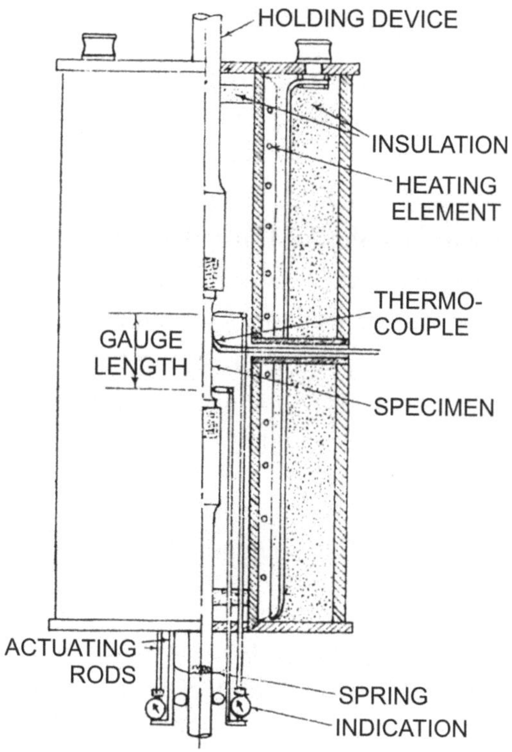

Fig. 6.4 Features of creep testing machine.

Many types of equipment are employed but their basic characteristics are the same. Depending on the design, the furnace accommodates one or more specimens at a time. Whatever may be the unit, the extensometers should be very sensitive and accurate.

While conducting the test, special attention should be paid to the tip of the thermocouple. It should make contact with the specimen under test.

According to the A.S.T.M., the temperature should not vary $\pm 3^{o}F$ from the average test temperature. The Load is applied directly and should be measured with an accuracy of 99%.

Standard Specimen

Round specimens of 16 mm, 12 mm and 6 mm (nominal) diameters with gauge lengths equal to four times the diameter are recommended. The larger sizes are preferred because measurement of strain will be easier and accurate.

During the test, the specimen is fixed in the machine and is heated to the temperature of the test in the unloaded condition. After the temperature of testing is attained by the specimen the load is applied; quickly, but without any shock. The initial instantaneous extension is noted, which consists largely of the elastic strain. There after measurements of strain are made at fixed time intervals. At least fifty observations are needed to plot the creep curve. At each observation of the strain the temperature also should be recorded. The average of all these temperature readings should be reported as the actual test temperature.

The Creep Testing Machine

The photograph of an Avery-Denison Creep testing machine (Model Type 48) is shown in Fig. 6.5. The machine has a four-column frame and a loading capacity of 3000 Kgf. The loading is accomplished by a load lever system with a leverage of 40: 1 so that, only a maximum load of 75 Kgf. is used actually. To maintain the perfect axiality of the loading, coplanar knife edged universal joints are provided in the loading system. The standard furnace is wound with Kanthal A in three separate zones and moves vertically on the machine columns to facilitate the specimen replacement. The control cabinet is mounted on the rear main columns and contains the furnace temperature control, variable rheostats for controlling the heat input to the three furnace zones, and an hourmeter to record the time taken by the specimen to rupture.

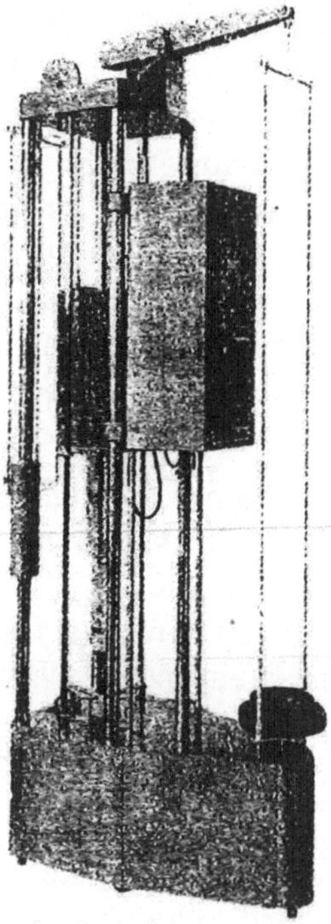

Fig. 6.5 Creep testing machine.
(*Courtesy* : Avery-Denison Ltd., Leeds, U.K.)

Standard specimens that can be used include those up to 16 mm (0.63″ approximately) diam. with gauge lengths of 75,100 or 125 mm, or 6 mm (0.25″ approximately) thick plate specimens. They are fixed with suitable grips made of a high temperature alloy (Nimonic 80A). An optical extensometer or an electronic extensometer (supplied as an accessory) reads the extension. Another important accessory is the controlled atmosphere heating chamber for conducting special tests.

The machine is also supplied as a constant stress rupture machine if required. In this, the standard 40:1 constant load lever sytem is replaced by a proportional weight lever/cam loading system. The capacity is only of 1000 Kgf., and utilizes a specimen of 25 mm gauge length. The cam is so designed that as the specimen extends, the lever ratio of 10:1 (initial) reduces proportionally with the cross-sectional area to maintain a constant stress on the specimen, over an extension of 12.5 mm (50%).

Interpretation of the Creep Data

Engineering members are designed to last through years and years of service. The number of years of service and the temperature at which they are to operate are the two factors which should be considered in the design. Sometimes the consideration of even these two is not sufficient in design. The permissible plastic strain during the period of service should also be taken into account. The material under consideration is tested, the creep curve is extrapolated to the required number of years on the X-axis and the strain is noted. Once it is less than the maximum permissible strain, the material is deemed to be satisfactory.

Because creep testing is a process which takes a very long time, it is common to test the metal up to the beginning or middle of the secondary creep only. The curve is then extrapolated to the required time period. This process is found to yield good results, for the strains so obtained are always more than the actual strains noted when the test is carried out until such time. The practice gives excess tolerance which is good from the point of view of a sound design.

Another short-cut method to ascertain the high temperature properties is to find out the hot-hardness of the material. It is established that the hot-hardness is linearly related to hot-strength (Fig. 6.6). This method is very useful, especially with brittle materials. However, no idea of the ductility of the material is obtained, which is sometimes absolutely necessary.

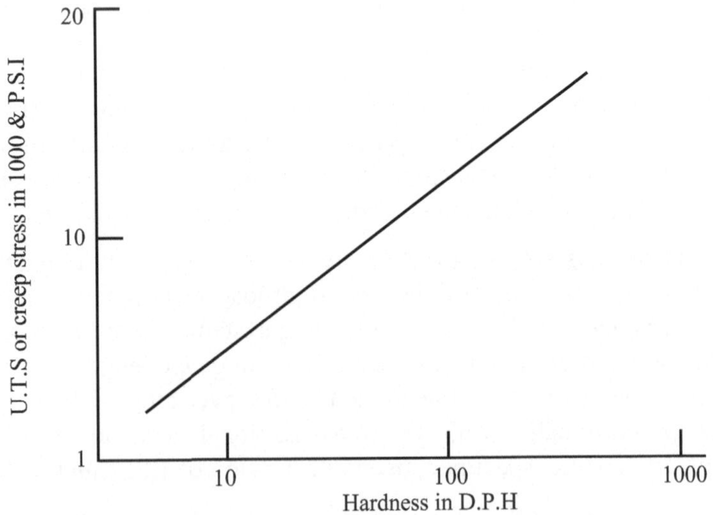

Fig. 6.6

FACTORS WHICH EFFECT CREEP

The following metallurgical factors influence creep properties of steels :

1. Composition
2. Heat treatment
3. Method of steel making
4. Grain size
5. Environment.

Alloying elements which go into solid solution into ferrite Mn, Ni, CO, etc., are found to improve creep properties of steels when the service temperatures do not exceed the recrystallization temperature of the steel. On the other hand carbide forming elements like Mo, W, V, etc., will promote creep properties at temperatures above recrystallization temperature range.

Optimum creep properties are derived when the steel is normalized. Annealed structures are next best. Quenched structures are very inferior.

Electric induction furnace steel is found to possess superior creep properties. Next best is the electric arc furnace steel, whereas, open hearth steel is the worst in this aspect.

A coarse grained steel is found to possess superior creep properties for service above the equi-cohesive temperature whereas a fine grained steel is preferred when operating temperature is below equi-cohesive temperature.

Creep tests on single crystals of zinc have revealed that creep practically stops when they are plated with copper. Again when the copper plating is electrolytically removed, they began to creep at the original rate. Further, the atmosphere in which the metal does service plays a great role. Oxidizing and corrosive atmospheres reduce the creep strength considerably.

SHORT TIME HIGH TEMPERATURE TESTS

Recent advanced technology warrants materials which are loaded and heated up in a matter of seconds. This is especially true in the case of missiles and rockets. Short-time high-temperature testing has thus become a must to ascertain the behaviour of the existing alloys and develop new alloys which withstand these conditions.

Test results are the most dependable when the test is performed under the actual service conditions. The testing consists of heating a special test specimen by its own resistance under a high load. The temperature and elongation are recorded by special automatic equipment. They are plotted to give a curve which resembles an ordinary stress-strain curve, the only difference being that instead of load as in the conventional tensile test, the temperature causes the failure in this case.

Sometimes the temperature required to effect a pre-determined amount of elongation is noted as the yield temperature. The yield temperatures are determined for a number of loads and a curve is plotted between the yield strength and the yield temperature. This curve is a good guide for ascertaining the behaviour of the material at different temperatures or loads (Fig. 6.7). However, it should be borne in mind that short-time high temperature test data is different from the creep data. Both should not be confused.

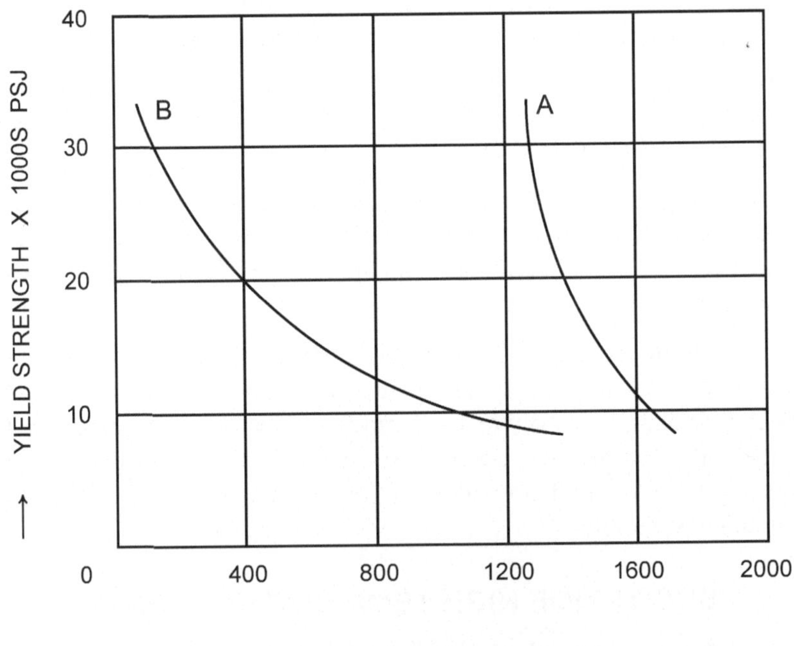

Fig. 6.7 Comparison of short-time high temperature (A) high heating rate (B) yield strength for an annealed 18/8 st. steel (heating rate 800° F/sec).

(Symposium on short-time high temperature testing, A.S.M., 1957)

Relationship between the Creep Rate, Stress and Temperature

It is established that the creep rate is dependent not only on the time of test/service but also on the stress applied/experienced by the metal. An empirical relationship governing the three variables is given by Baunamann as follows :

$$C = f / E.\ T.$$

where C = creep rate

f = stress at the particular temperature of test

E = modulus of elasticity at the particular temperature of test.

and T = time required to obtain creep equal to the elastic deformation.

HIGH TEMPERATURE MATERIALS

High temperature alloys are developed gradually. Of course, their development was a result of the prevalent needs. In the beginning, alloys suitable for use in steam engines and steam power plants were developed. These were mostly ferritic steels to which chromium and molybdenum were added. These form carbides which disperse throughout the alloy lattices and prevent softening. Typical alloys of this type consist of 0. 5 % Mo, 0.1 to 0.3% and 1.25 to 5.0% Cr. These alloys are suitable only up to temperatures of about 600°C. At high temperatures, phase changes occur and the alloys soften.

Later advancements with the advent of jet engines are the austenitic alloys. Nickel based and nickel chromium complex alloys and cobalt based alloys are developed to work up to about 750 °C. In all these alloys the percentage of carbon is very carefully controlled, for if it exceeds an optimum limit, the carbides formed will coalesce instead of being present as finely dispersed particles. The rupture strength is lowered. The austenitic alloys consist of carbon up to 0.1 % (max.), 12 to 25% Ni, 15 to 18% Cr with traces of Cb, Ti, and/or aluminium. These trace additions increase the strength of the alloys by forming compounds with nickel (Ni_3Ti and Ni_3Al) which are stable and hard and get distributed finely throughout the structure. These alloys are harder. The following are some nickel-based high temperature alloys.*

	C	*Cr*	*Ni*	*Mo*	*Co*	*Ti*	*Al*
Inconel	0.04	15.5	76.0	–	–	–	–
Inconel X	0.04	15.0	75.0	–	–	2.5	0.6
Nimonic 90	0.08	20.0	58.0	–	16.0	2.3	1.4
Hastelloy B	0.1	1.0	65.0	28.0	–	–	–
Rene 41	0.1	19.0	53.0	10.0	11.0	3.2	1.6
Udimet 500	0.1	19.4	55.6	4.0	14.0	2.9	2.9

*H.C. Cross and W.F. Simons–Alloys and their properties for elevated Temperature Service. "Utilization of Heat-resisting alloys". A.S.M., 1954.

Some of the cobalt-base alloys are as follows:

	C	*Cr*	*Ni*	*Mo*	*Co*	*W*
Vitallium (HS-21)	0.25	27.0	3.0	5.0	62.0	–
X – 40 (HS-31)	0.4	25.0	10.0	–	55.0	8.0

Some complex superalloy compositions are as follows:

	C	*Cr*	*Ni*	*Mo*	*Co*	*W*	*Cb*	*Ti*	*Al*	
N–155	0.15	21.0	20.0	3.0	20.0	2.5	1.0	–	–	0.15 N
S–590	0.4	20.0	20.0	4.0	20.0	4.0	4.0	–	–	
S–816	0.4	20.0	20.0	4.0	Balance	4.0	4.0	–	–	3.0 Fe.
K–42B	0.05	18.0	43.0	–	22.0	–	–	2.5	0.2	
Refract aloy 26	0.05	18.0	37.0	3.0	22.0	–	–	2.8	0.2	18.0 Fe.

Note : All the above alloy compositions consist of iron as the balance percentage, except S - 816.

The latest class of the dispersion strengthened high temperature alloys consist of Al_2O_3, ZrO_2 and SiO_2. These oxides are introduced artificially into the alloy matrix to function as the thermally stable phases. Powder metallurgy comes to the rescue of the metallurgist here. After sintering, the alloy is extruded into the required shape.

Based upon the sintering technique, another stride has been made in the development of high temperature alloys. Alloys called cermets are developed by blending ceramic powders such as barites, silicides and carbides with a metal binder. Alloys which stand up to 900 °C have been successfully developed by this technique. Still there is a long way to go in the development of high temperature alloys.

References

Johnson, A.E., Henderson, J. and Khan, B., Complex stress creep relaxation and fracture of Metallic alloys, H.M.S.O., 1962.

Underwood, E.E., J. Inst. of Metals, Vol. 88, pp. 266-271, 1959-60.

Underwood, E.E., Materials and Methods, Vol. 45, pp. 127-129, 1957.

Creep and Recovery, A.S.M., Metals Park, Ohio, 1957.

Garafalo, F., Smith, G.V. and Royle, R., Trans. A.S.M.E., Vol. 78, pp. 1423-1434, 1956.

Sully, A.H., Recent advances in knowledge concerning the process of creep in metals, Progress in Metal Physics, Vol. 6, Pergamon Press, London, 1956.

IS : 3407-1965, Interrupted creep testing of steel at elevated temperatures.

IS : 3408–1965, Non-intercepted creep testing of steel at elevated temperatures.

IS : 3409-1965, Creep stress rupture testing at elevated temperatures.

CHAPTER 7

Testing of Miscellaneous Products

(Testing of sheet, Strip, Tubing, Bearings and Welds and Spark Testing of Steel)

TESTING OF SHEET, STRIP AND PIPES

Materials in the form of plate, strip, tube, etc., are not easy to use as standard test specimens for performing the conventional tests because their thicknesses are small. Special procedures are adopted for determining the strength of plate specimens. Further, most of the sheet metal is employed in various fabrication processes. In these, the ductility is as important as the strength. Though the percentage elongation obtained in the tensile test gives us an idea of the ductility, it fails to be comprehensive in actual engineering fabrication practice. Hence it becomes necessary to test the sheet metal for "bending" and "cupping" actions in order to obtain information regarding its behaviour in actual fabrication. These tests, though arbitrary, provide a very useful basis in metal working because the conditions of testing, generally, approximate the actual 'working' conditions.

Testing of sheet, strip, plate, wire and tubular products presents many practical difficulties. The foremost of these difficulties is a standard specimen. Even when some standard specimen is used, the accompanying problem is gripping of the specimen in the testing machine. Ball or pin type grips and wedge grips can be employed for sheet and strip, wires and tubes, respectively. In the case of thin wires, wedge grips also are not satisfactory. In such cases the wire/wires under test is/are cast into an artificial lead grip and then tested.

In the case of wire rope, the tensile strength cannot be taken as the sum total of the strengths of the individual wires. The cold work, and consequent hardening, introduced during the process of stranding increases the strength of rope considerably. However, if the strength is taken as the sum total of that of the individual wires, it is better for design, for the actual strength will be definitely more and this extra serves as a tolerance.

Tensile Strength and Hardness

Standard specimens for the purpose of tensile testing of sheet and plate are shown in Figs. 7.1 and 7.2.

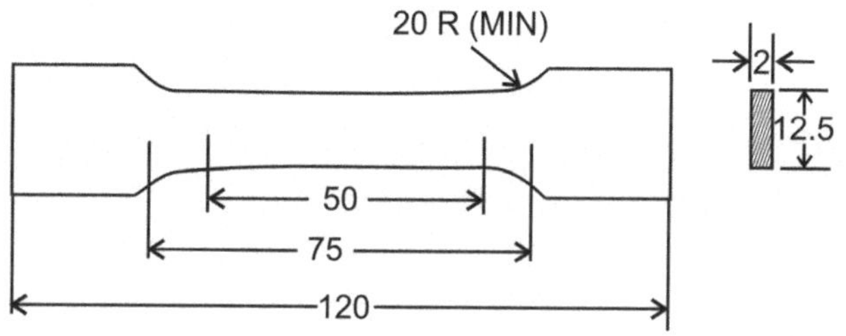

Fig. 7.1 Tensile test specimen (for sheet under 3 mm thickness). (All dimensions in mm.)

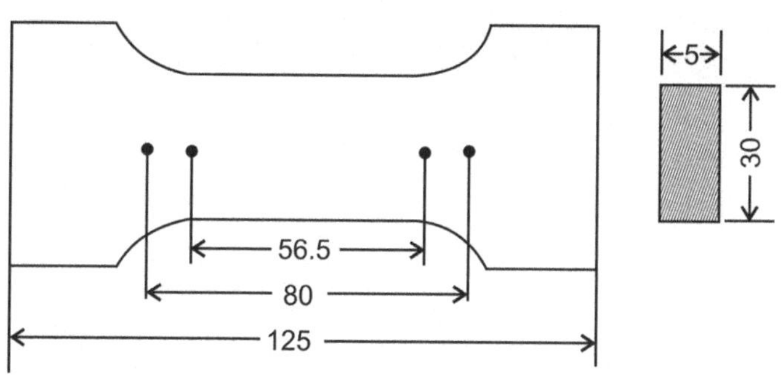

Fig. 7.2 Test specimen for tensile testing for sheet thicker than 3 mm. (All dimensions in mm.)

The tester should employ his discretion in the selection of the testing machine. There is no point in using a 20 T capacity universal testing machine for a sheet specimen whose thickness is 5 mm. If heavy duty machines are used for tiny tests the elongation or strain readings will not be accurate owing to the lack of sensitivity. A Hounsfield tensometer is an ideal machine for such a purpose. It is a small bench type machine which is operated by a hard lever for loading. The specimens, generally of sheet and plate, are gripped by pin type grips which slide on horizontal rods during loading. The stress values can be read by the movement of a mercury column over a graduated scale and the strain by means of a vernier. There will be provision for plotting the stress-strain diagram also during the test.

Once the specimen is tested onto breaking, calculation of the tensile strength is by the usual procedure.

$$\text{T.S.} = \frac{\text{Maximum load}}{\text{Cross-sectional area of the specimen}}.$$

Similarly, hardness values, if taken, should be measured by the Rockwell tester and not on a Brinell hardness tester. The elementary principle that the backing material behind the indentation in hardness measurement should be at least five times the diameter of the indentater, should not be ignored.

Ductility Tests

Bend test

This is a simple test which indicates the behaviour of the metals under bending loads (Fig. 7.3). The metal piece under test is fixed in a vice vertically and bent through an angle of 90°. A hand hammer (a wooden one) is the implement employed.

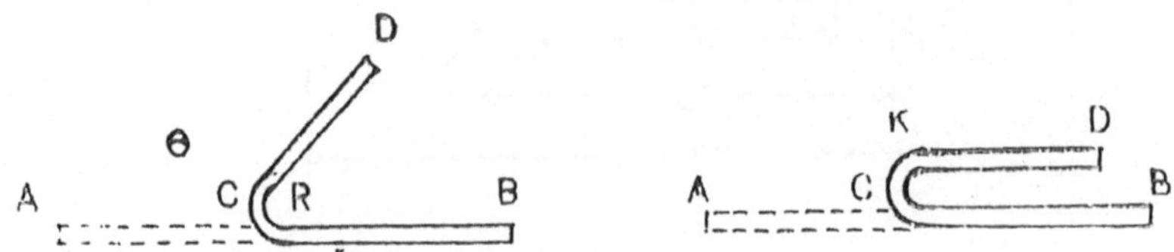

Bend test through a specified angle θ. Bend test through 180°.

Fig. 7.3 Bend test.

Some designers specify that the bending should be continued further. In such cases, the sheet is taken out of the vice and hammered as to double it. Thus it is bent through an angle of 180° (Figs. 7.4 and 7.5).

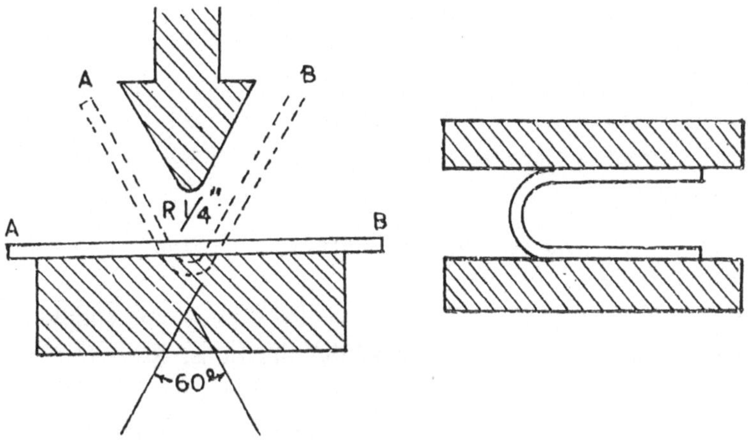

Fig. 7.4 Bend test for plates.

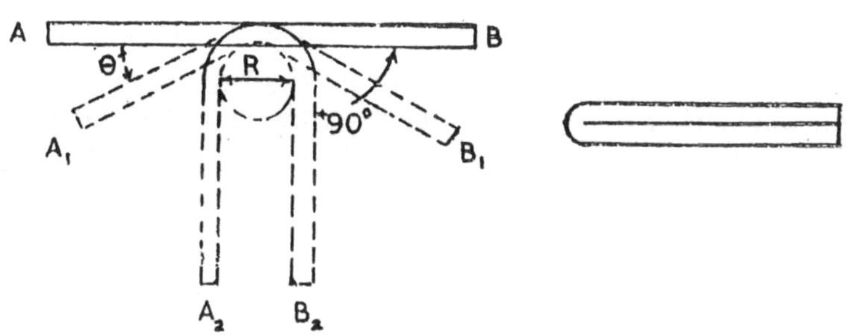

Fig. 7.5 Bend test.

Reversed bend test

This is a more drastic test than the bend test. The specimen, after being bent, is straightened and again bent in the opposite direction. Instead of performing the test with a vice and hammer, if it is conducted in a sheet metal press, the magnitude of the bending load can also be measured.

The above tests serve only as guides, in addition to the conventional tensile test data regarding the percentage elongation. The ease with which the sheet or plate is bent or reverse bent gives us an idea about the work-hardening tendency of the metal, in addition to its formability.

Pressing properties

When sheet metal is to be employed for pressing, deep pressing or spinning, data regarding its behaviour under similar conditions is essential. Simple percentage elongation and bending properties are not sufficient.

In press work and spinning the stresses involved are both tensile and compressive varying in magnitude and direction according to the final shape of the article. For satisfactory results, the material should possess the same percentage elongation in all the directions, i.e., both along, and at right angles to, the direction of rolling. It is further desirable that the material should record the same percentage elongation when tested on either 2″ or 8″ gauge length specimens.

Another important requirement is that there should not be much difference between the yield stress or 0.1% proof-stress and the tensile strength of the metal. This means that the plastic portion of the stress-strain curve, in which region most of the pressing operation is performed, should be fairly large. Further, the shape of the curve is also important as it indicates the rate of work-hardening. A curve which tends to deviate from the straight line elastic position for small increases in stress means a lower rate of work-hardening. Such a metal gets work-hardened easily.

The grain size of the metal is very important from the pressing point of view. Though a fine grain structure may not give the same ductility as a coarse structure, the former is preferred because it does not fail by splitting due to the relative movement of the die/punch and the sheet surfaces.

Lastly, in deep drawing processes the compression of the metal should also be considered. The material under compression between the punch and the die should not work-harden fast. Simultaneously the tensile stresses should not become excessive. That is why the sheet in an annealed condition is best suited for deep pressings than in other heat treated conditions.

Cupping Test

Many testing machines–Erichsen, Olsen, Amsler, Avery, etc., are available. The operating principle in all these machines is the same, viz., indenting the sheet metal with a hemispherical plunger and noting the depth of indentation, i.e., depth of the cup formed before failure occurs. The only difference will be the dimensions of the plunger head.

Erichsen cupping testing machine

The principle is illustrated in Fig. 7.6 as a line diagram. The sheet usually a 89 mm square is held between two ring clamps. A punch with a hemispherical face is forced against one side of the sheet. The dome produced is measured in millimetres while the failure of the metal can be observed in a mirror attached to the cup side of the machine.

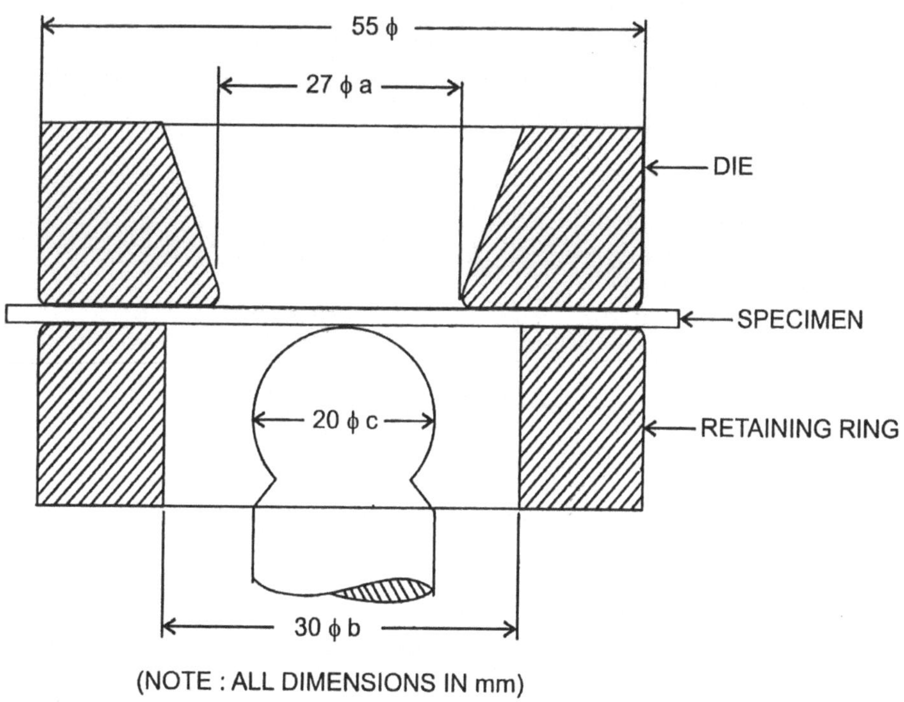

Fig. 7.6 Erichsen cupping test (modified).

In Olsen test the depth of the cup is reported in thousandths of an inch. Some modern machines are equipped to measure the force applied to the punch.

Variables

Some differences in test results may occur due to the following variables:

1. Variation in the clamping pressure.
2. Variation in the rate of loading.
3. Variation in the nature and quantity of the lubricant applied to the sheet specimen.

In spite of the above variations and the otherwise established scatter in the test results on the same batch of material, the cupping test is still widely employed. It needs no elaborate specimen preparation, and is rapid. The general condition of the specimen is readily observed. The development of surface roughness, the failure depth and the load provide a useful guidance as to whether the material is useful for a particular drawing operation or not.

The dies and the indenting tool dimensions for different testing machines are given in the table below :

Type	*a*	*b*	*c*
Erichsen	27	–	20
Avery	27	–	20
Olsen	–	–	–
Amsler	50	–	20

(All dimensions in mm)

Deep Drawing Tests

Deep drawing is one of the common fabrication processes. A study of the behaviour of metals under conditions of deep drawing has thus become necessary. Cupping tests have only a limited drawing process in them and as such the data obtained in such a test does not give much understanding regarding the behaviour of the metal under conditions of actual deep drawing. Deep drawing tests are evolved which are essentially a modification of the cupping test.

The machine is an autographic press designed to produce a cup of 2″ (5 cm) internal diameter, from a circular metal blank. A case- hardened steel punch of circular diameter is used for applying the load. The punch has a flat surface with a 1/4″ (6 mm approximately) round edge to facilitate its sliding freely into the cast iron cylindrical die of $2\frac{1}{8}''$ diameter (50 mm). The 1/16″ (1.5 mm) clearance is for accommodating the sheet. The punch load can be measured and recorded.

The blank holder pressure is important to avoid 'slipping'. Pressures as high as 10,000 lbs (5000 Kgsf. approximately) are employed for gripping the blank.

The general form of the autographic diagram is illustrated in Fig. 7.7. These diagrams are similar in appearance for all the materials though the loads vary. The initial rise of the load is fairly sharp, the maximum being reached immediately after the walls of the job become straight (C). Then the loads remain steady at the maximum value for a considerable period of the draw and, subsequently, fall. This fall is rather rapid.

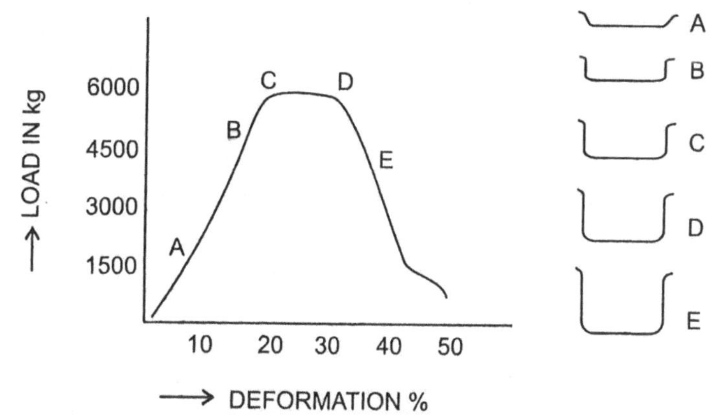

Fig. 7.7 Deep-drawing test on aluminium. (Autographic diagram.)

The maximum load (top) portion of the curve approaches a rounded shape when the drawing is rapid or the lubrication conditions are favourable. The load starts to fall once the cup rim reaches the radius of the die.

Precautions in conducting the Cupping Test

1. The test blank should be machined on its outer edge to ensure accuracy in dimensions.
2. Positive blank holder clearances are not advisable for any blank thickness.
3. Graphite lubrication is recommended to minimize the effects of large drawing speeds and friction.
4. The die profile should be liberally curved in order to minimize the friction.
5. The punch surface also should be generously curved so that small irregularities in the surface of the specimen do not effect the drawing.
6. Sufficient radial clearance should be provided between the punch and die; except during the final stages of drawing.
7. The measurement of the cup depth should be made, as far as possible, on a scale magnified to the maximum. This ensures greater accuracy in measurement.

Cupping test *vs* the Drawing test

Elaborate experiments have led to the following distinction between cupping and drawing tests:

The cupping test simulates the operations wherein stretching is resisted by externally imposed constraints. In the drawing test, the stretching which occurs is induced due to the resistance of the metal to lateral drawing.

TESTING OF PIPE AND TUBULAR PRODUCTS AND WELDS

Before proceeding to the testing of tubes we should have a clear idea regarding the limitations of the word "tube". Many confuse tube with pipe. In fact the difference between them is so insignificant that their meanings overlap. Broadly, we may define the tube as that article of circular cross section whose wall thickness is negligible compared to its overall diameter. On the other hand, the wall thickness of pipe will be a noticeable fraction of the overall diameter. For practical purposes the author feels that the limiting ratio (wall thickness to the overall diameter) can be regarded as 1:10.

Testing of Tubes and Pipes

Tension test

The tensile strength of pipe and tubing is not of much significance. However, it can be determined in the same way as for other items. The standard test specimen is illustrated in Fig. 7.8.

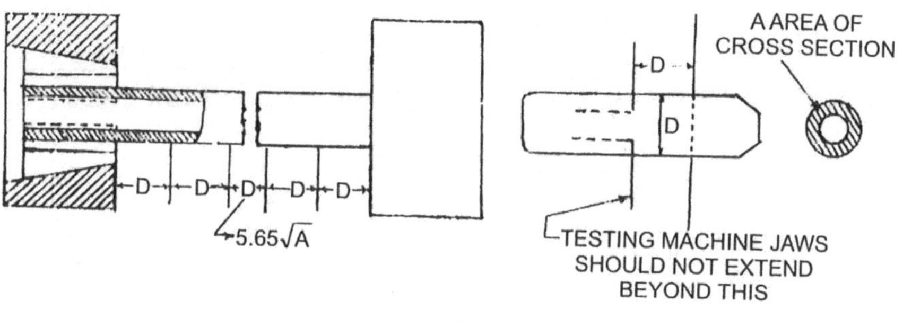

Fig. 7.8 Standard tension specimen and fixture for metal tubes

(up to 2″ O.D.) A.S.T.M. (From *A.S.M. Metals handbook*, 1948.)

One important problem which arises in this case is gripping the specimen in the testing machine. One method is to flatten the two ends with a hand hammer and grip them as if it were a plate. Sometimes threads (internal or external) may be cut and the two ends can be screwed to the fixtures containing threads, specially made for the purpose.

Whichever method is adopted, the gauge length portion should not be subjected to any deformation before the test.

The actual material cross-sectional area (shown shaded) should be computed and used in calculating the tensile strength (Fig. 7.9). It is recommended that this value should be the average of two or three values taken at different places on the gauge length.

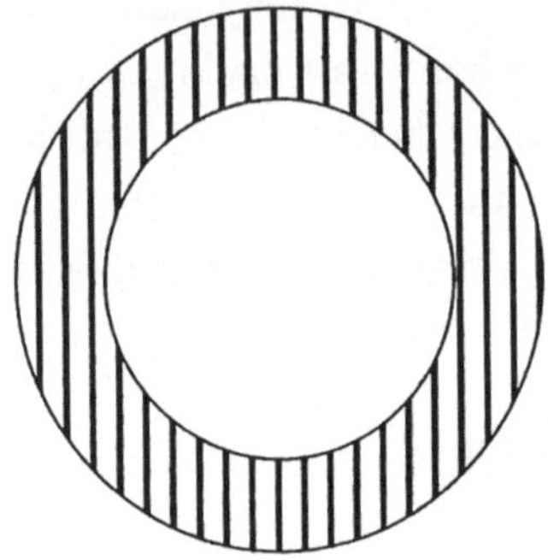

Fig. 7.9 Cross-section of a pipe. (Wall-thickness shown shaded.)

Testing of Welds

With the development of welding technology, all the nut and bolt and riveted joint designs are being done away with and welding has become an indispensable process of metal joining. Nowadays welding can be performed on any metal or alloy. Ship building, tankers, pressure vessels and construction machinery are a few fields that can be cited to remind the reader of the importance of welding. Further, the molten weld metal is used to repair and rebuild minute process damages by filling the nicks or cavities produced, especially in rolling mill rolls, camshafts, etc.

The mechanical properties frequently determined in the case of welded joints are the strength and ductility.

Tension test

The welded joint is tested for tensile strength in the same manner as any other metal object is tested. A standard specimen of 2″ (50 mm) gauge length is shown in Fig. 7.10.

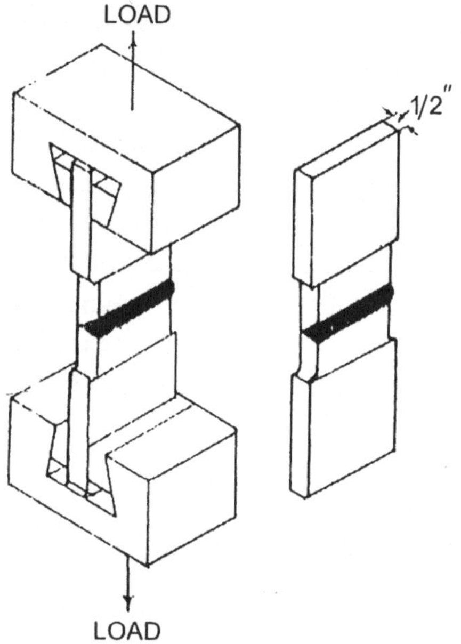

Fig. 7.10 Tensile test specimen and fixture for welds.
(From *A.S.M. Metals Handbook*, 1948.)

This test suffers from certain drawbacks. The specimen being made out of the base plate as well as of the weld metal, gives readings which do not exactly record the properties of the weld proper. When welding is performed with a high strength welding rod or electrode, the yield point recorded will be that of the weld metal whereas the percentage elongation obtained will be that of the plates joined. In the case of welds which are weak in comparison with the plates, the yield point recorded will be that of the plates and the elongation of the weld metal. Thus, in order to obtain a reliable strength data it is advisable to obtain a longitudinal specimen along the weld and perform the test. A 0.505″ (13 mm) diam. standard specimen can be obtained from a 5/8″ (15 mm) plate weld and 3/8″ (10 mm) diam. specimen from a 1/2″ (12 mm) plate weld with lengths of 2″ (50 mm) and 1.5″ (38 mm),* respectively.

*Approximate values.

In the testing of butt and fillet welds, a bend test or a guided bend test is employed to measure the soundness and ductility. Fig. 7.11 shows the design and the method of preparation of the standard specimen as well as the sequence of operations in conducting the test. Plate welds with more than 1/2″ (12 mm) thickness need heavy presses. Hammer action is not recommended, for the action is that of impact. The specimen is bent until failure occurs. Corner cracks or any surface failure less than 1/16″ (1.5 mm) dimensions are not considered as failures. The elongation of the outside fibres is directly measured with the help of a flexible scale. The ductility can also be measured with the help of a bend extensometer.

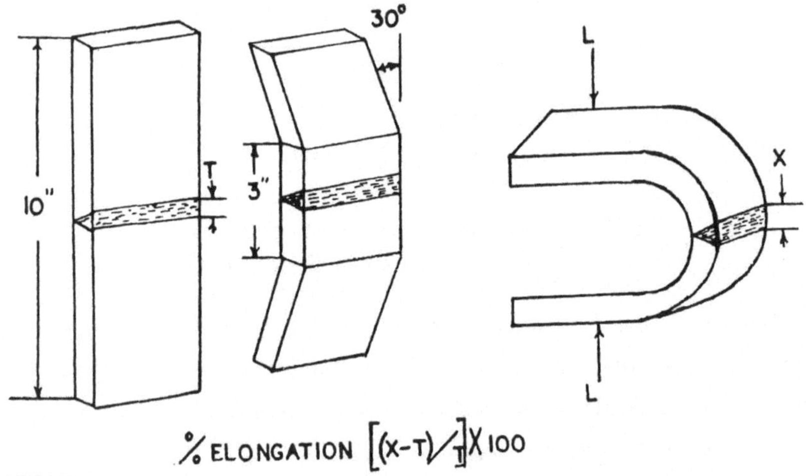

Fig. 7.11 The bend test. (From *A.S.M. Metals Handbook*, 1948.)

A guided bend test, as it is called, is sometimes employed. This is a modification over the aforesaid bend test. The bending action and thereby the failure is made to occur within the gauge length, by ensuring the bending operation between formers, a male plunger, while the plate under test acts as a female member between two grips.

Another test called the root bend test is performed to examine the penetration of the weld metal. This test is performed by bending the welded plate backward. This action stretches the bottom of the V of a single Vee joint.

A side bend test is similar to the root bend test except that the specimen is taken transversely across the length or plane of the weld (see Fig. 7.12).

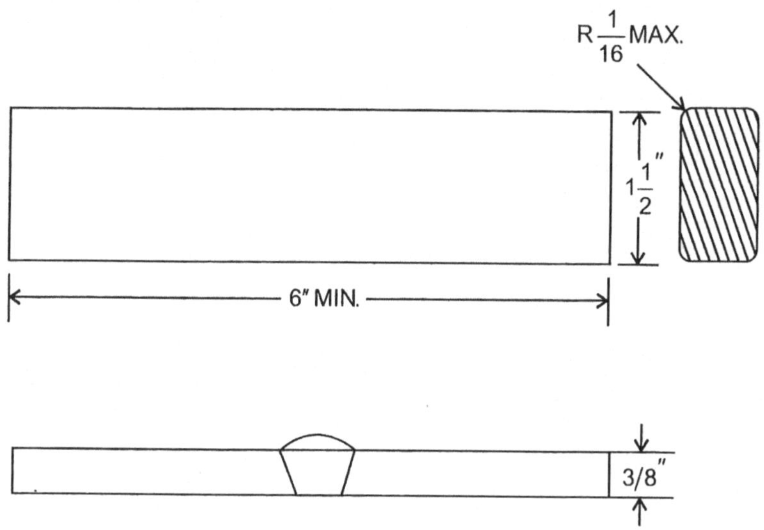

Fig. 7.12 Specimen for face and root bend tests.
(From *A.S.M. Metals Handbook,* 1948.)

Tensile impact

A tensile impact test is made on the welded specimen, just as the tensile test is made. This test reveals the capacity of the welded joint to resist impact loads in tension.

The nick-break test

The method of this test is illustrated in Fig. 7.13. This is a severe test and is very useful in revealing the homogeneity of the weld metal. Further, it is very simple to perform. It consists of rupturing with a hammer blow, a specimen made out of welded plates and notched at both the ends of the weld region. By examining the fracture surface we can see whether the weld penetration is proper and the weld is free from inclusions like slag, oxide, etc.

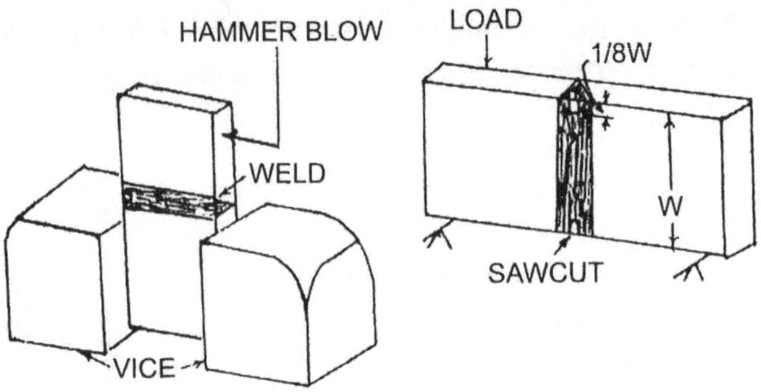

Fig. 7.13 Nick-break test. (From *A.S.M. Metals Handbook*, 1948.)

The dimensions of the A.S.M.E. unfired pressure vessel code for paragraph U 69 quality vessel specimen are as hereunder.

Width – $1\frac{1}{2}\ T + 1/2''$ $(1\frac{1}{2}\ T + 12\text{mm})$*

Slots – 1/4″ (6 mm)* deep at both ends of the weld (with a hacksaw)

where T is the thickness of the plate.

Modified nick-break test

The nick-break test is modified to suit the testing of fillet welds. A specimen is made by welding two plates at right angles to each other. A pressure or blow is applied on the welded edge so that the weld is opened up. The fractured surfaces reveal lack of penetration, inclusions like slag, oxide, etc., if any (Fig. 7.14).

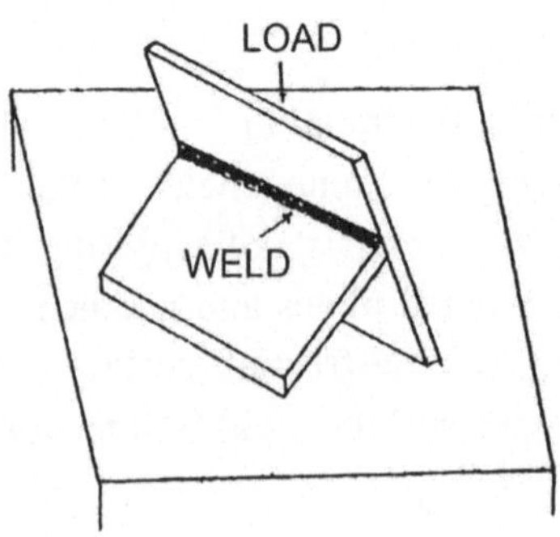

Fig. 7.14 Fillet weld break test. (From *A.S.M. Metals Handbook,* 1948.)

*Approximate conversions

Non-destructive testing

Many a time, it becomes necessary not to spoil the welding made and at the same time ascertain its soundness. X-ray radiography is a common method of non-destructive-weld examination. How the various defects are revealed in an X-ray radiograph is listed in the chapter on Radiography elsewhere in this book.

Tests on welded pipe and tube

Welded pipe and tubing are tested for different properties as follows. Most of the tests are 'to be specified' tests.

The drift test

It is very difficult, rather impossible, to obtain a tensile specimen from a thin walled tubing. The drift test is devised to determine the strength of the weld. It consists of forcing a tapered hardened steel drift into one end of the tube which is welded longitudinally. The drift is preferably pressed. The end of the tube is gradually expanded until it splits, usually outside the welded area, showing that the weld is stronger than the base metal (Fig. 7.15). The distance through which the drift has entered into the tube may be taken as a measure of the ductility of the welded tube. It should be noted that in this test, the welded joint is stretched, an action which is similar to that in a tensile test.

The crushing test

This test is a 'to be specified' test. A specified length of the specimen is cut and compressed longitudinally to the prescribed size (Fig. 7.16). During the compression, it should not register any crack or break.

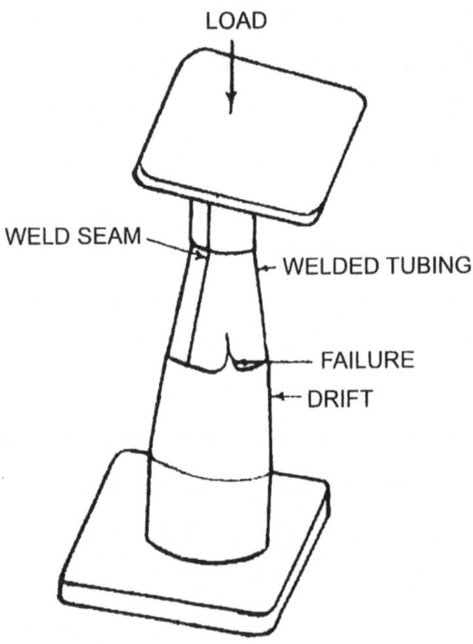

Fig. 7.15 Drift test.

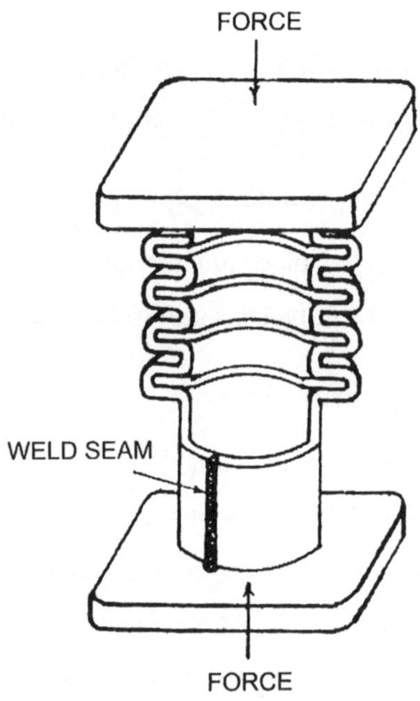

Fig. 7.16 Compression testing of welded tubing.

(From *A.S.M. Metals Handbook*, 1948.)

Flange test

This test is made on short lengths cut from the tube. A flange is made at one end using the flaring tool and a die block of specified dimensions. The dimensions are given in Fig. 7.17.

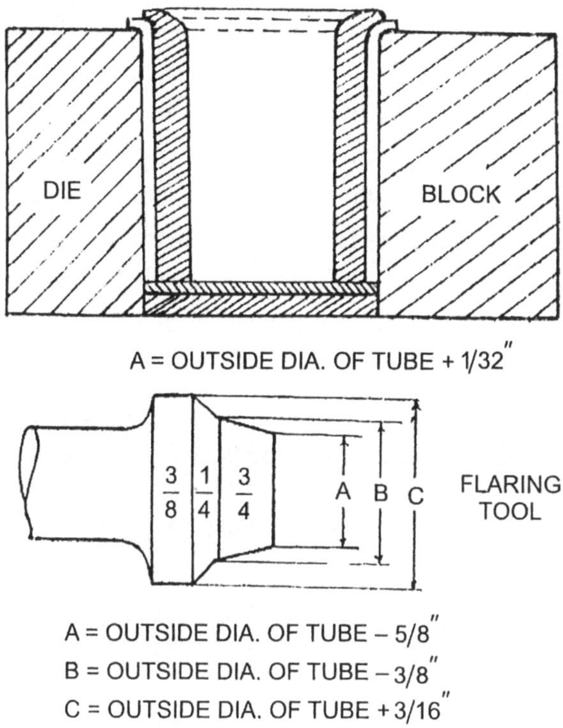

Fig. 7.17 Flaring tool and die block for flange test.

(From *A.S.M. Metals Handbook*, 1948.)

Pressure and leak tests

This test is a 'to be specified' test. The welded job–a pressure vessel or pipe is tested to operate without any leak at a pressure of

$$X + \left(\frac{x}{100}.X\right)$$

where X = operating pressure, and

x = percentage excess pressure specified by the indenters.

Either pneumatic or hydraulic pressure is employed in the test but the latter is common. However for routine testing of pipes, the testing pressure is not more than that described by Barlow's formula:

$$P = 2\ St / D$$

where P = Pressure in p.s.i. (or Kgf/mm^2),

S = Maximum allowable fibre strength p.s.i. (or Kgf / mm^2),

t = thickness of the metal pipe in inches (or mm), and

D = outside diameter of the pipe in inches (or mm),

Further, bursting test–a test in which the pressure is gradually increased until the component bursts–is also conducted. In some instances the test is conducted until a certain pressure value (to be specified) and the permanent set is measured. In most cases, the permanent set test is conducted by enclosing the pressure apparatus in a jacket containing a known volume of water. Unless otherwise specified the permanent set tolerance is 10%.

TESTING OF BEARING MATERIALS

There are many types of bearing materials in use. The common ones are based on lead, tin, cadmium, iron and copper. The latter two, viz., the copper-and iron-based bearing materials, are also made by the powder metallurgical process.

The bearings can neither be decisively classified nor their properties listed as important or unimportant, for the bearing has to withstand a variety of imposed conditions. The relative importance of the properties of the bearing will change depending on the application, i.e., the machine, part it is used in. Thus, the choice of the material is governed by the service conditions. Hereunder, we consider some important mechanical properties and the methods of their determination.

Mechanical Properties and Methods of Their Determination

Resistance to fatigue

The fatigue resistance of a bearing material is tested only in actual service. Preparing a laboratory sample out of the material and testing it as an ordinary steel sample does not give us the exact life of the bearing because resistance to bearing fatigue is a distinctive property other than the ordinary fatigue. The temperature developed in actual service is a vital factor which makes all the difference. It is established that fatigue cracks originate from the shaft contact surface and propagate down to the bond surface between the bearing and the steel backing.

Antiseizure characteristics

When the bearing is running at a very high speed or stationary under a heavy load, or when the lubricant film is not properly spread between the shaft and the bearing surface, the surfaces of the bearing and shaft come into actual contact. The rubbing action produces heat and the temperature is raised. Sometimes the temperature becomes so high that the two surfaces get 'welded'. Such an extreme case is referred to as "seizure". It is noted that some materials withstand these conditions better than others. Such alloys are said to possess better anti-seizure properties. Popular bearing materials are in the following order of anti-seizure characteristics–white metals, copper-lead, leaded brozne and bronzes.

No test for this property is specified except a knowledge of lubrication and the melting points of the metals in contact.

Bonding ability

If the bearing bush is not properly bonded with the bearing, it will start to give jerks and will gradually chip or peel off from the backing. For this property also, there is no specified standard test but the manufacturers will furnish the required instructions along with the supplies.

Compressive strength

This is a vital factor in the bearing life. If the compressive strength of the bearing is not sufficiently high, it will extrude under the load. It is a general phenomenon with metals that they become effectively stronger, the shorter they are. Hence, the recent trend is to reduce the bearing thickness as much as possible in order to increase its effective compressive strength.

The compression test is generally performed by taking a 1" diam. (25 mm)* and 1" (25 mm) long cylinder and loading it until it registers a 0.001** compression strain. The value of the load in lbs. (or kgf) is reported as the compression strength.

An interesting point which should be noted here is the increase of the deformation with time in the case of babbits. This makes a lot of difference in the actual performance and the values found in testing, especially with respect to hardness, compression and fatigue. However, the observed values need not, and cannot, be overlooked as they serve as a useful guide both for manufacture and in design.

---

* Approximate values.

** This value may be specified as is desired (to suit the requirement).

Compressive strength at breakage, sometimes called the crushing strength, is calculated from the following empirical formula:

$$P = KWT^2 (D - T)$$

where P = radial crushing strength (prescribed as a minimum),

W = bearing length in mm,

T = wall thickness in mm,

D = outside diameter of the bearing in mm, and

K = strength constant in Kgf/mm^2 (varies between 10–30 depending upon the material).

(The value of '*K*' is about 12 for copper-based and 15 for iron-based powder metallurgical bearings.)

This test is also a working condition test. The strength is determined by increasing the shaft load on the bearing assembly and noting the load at which the bearing actually gives way. The above formula gives the theoretical value to be the minimum radial crushing strength. The actual crushing strength should be more than this value.

Deformability

This is a useful property especially in cases where the bearing is used on flexible shafts. Under loads, the shafts deflect at a considerable angle when going through the bearing. Such a condition causes heavy pressures on the bearings and the lubricant film may rupture leading to seizure or bearing failures. If the bearing has a good deformability it takes care of itself and aligns properly. This is useful in fittings like electric motor, auto engines, etc.

The material is tested under conventional stress conditions and the yield point noted. If the yield point recorded is lower, it is supposed to possess a good deformability. The yield point should be lower than the bearing loads, of course with no adverse effect on the overall strength.

SPARK TESTING OF STEEL

We know that ancient man made fire from the sparks of stones. He rubbed two stones to produce a spark. Spark can be understood as the flow of particles, thrown away from an object, when it is subjected to abrasion and/or impact. When two objects are rubbed against each other, tiny fragments are delivered out. These burn in the atmospheric air and exhibit a glow.

A piece of steel or cast iron, when held against an abrasive wheel, gives out a stream of sparks. These sparks are, as we have already known, tiny fragments of the material itself. These tiny fragments also contain carbon as does the parent material. Further, these pieces are at an elevated temperature due to the friction. The high temperature and exposure to the atmospheric oxygen makes the carbon burn. We see the glow.

If we consider a plain iron carbon alloy, the carbon is existing either as iron carbide or both as a mixture of iron carbide and free graphite. But when the alloy contains some other elements which directly or indirectly influence the existence of carbon, the mode of the sparking is also influenced. Conversely, the nature and appearance of the spark is based on the composition of the alloy. This is methodically studied and a broad guide is evolved which leads to the prediction of the composition of the steel by a close study of this spark stream. This technique is termed as spark testing of steel.

Terminology

When the steel is held against a revolving abrasive wheel, we see that tiny particles of the steel are separated and are hurled through the air. The trajectory of these particles, a glowing streak, is called the "carrier line". The whole picture as we see, emanating from the place of contact of the steel and the abrasive wheel, is called the "spark picture". Sometimes, the spark bursts, giving a flowery appearance to the carrier line. A volume increase occurs the moment the carbon is oxidized to carbon monoxide. The steel speck, though very hot and plastic, cannot accommodate such a sudden increase in volume and 'bursts'. These are called 'carbon bursts'. Another important phenomenon is "forking", when the carrier line appears to be the stem with smaller carrier lines emanating from it, the whole thing giving an appearance of a winter tree or blades of grass. A large number of forks which are not perfect forks themselves, appear on a few occasions. They are called sprigs. These and the main carrier line itself should be carefully observed for their colour, intensity, thickness, etc., during spark testing.

Testing Procedure–Appliances

The test procedure is very simple in that all that is required is to produce sparks from the steel under examination. Yet to obviate the human error, a certain code is laid down regarding the procedure. The grinding wheel should be made of carbo-emery with a minimum diameter of $1\frac{1}{2}''$ (38 mm) and 1/4″ (6 mm)* thickness. Further the shaft speed shall be 13,000–15,000 rpm [corresponding to about 4,500 ft (1,500 meters) per minute peripheral velocity]. The pressure applied should be such that the spark stream produced would be about 12″ (300 mm) in length.

It is advisable to isolate the spark testing chamber from brilliant illumination in order to facilitate a thorough examination of the spark picture.

The spark testing may also be performed by the use of a portable grinder. In such cases the total weight of the grinder should be about 7 lbs (3 Kgf.), well distributed. The grade of the emery used should be coarse corresponding to wheel No. 46 PV (Norton). This is desirable because the wheel should be able to cut out fragments from the steel under test.

Appearance of the Spark Picture

For this purpose we shall classify the steels into two common groups: Plain carbon steels and alloy steels.

Plain carbon steels

These steels do not contain alloying elements (except carbon) to any appreciable extent. These steels can be further classified as low carbon (up to 0.3% C), medium carbon (0.3–0.55% C), and high carbon (0.55 %C and above), steels. The spark pictures appear as follows.

Low carbon steels : The spark picture will be one of the simple forking (Fig. 7.18).

Medium carbon steels : The spark picture consists of both forking and some secondary bursts (Fig. 7.19). The higher carbon content is responsible for these bursts. The incandescent steel particle is disintegrated by the CO or CO_2 formed by the reaction of atmospheric oxygen with the carbon.

---

*Values in the brackets are approximate conversions.

Fig. 7.18 S.A.E. 1015 (C 0.14, Mn 0.14).
(*Courtesy* : Iron Age, Radnor, Pennsylvania, U.S.A.)

Fig. 7.19 S.A.E. 1030 (C 0.32, Mn 0.72).
(*Courtesy* : Iron Age, Radnor, Pennsylvania, U.S.A.)

High carbon steels: The spark picture will be similar to the one noticed in the case of medium carbon steels but the intensity of carbon bursts will be more due to increased carbon content (Fig. 7.20).

Fig. 7.20 S.A.E. 1095 (C 1.01, Mn 0.49).
(*Courtesy* : Iron Age, Radnor, Pennsylvania, U.S.A.)

Alloy steels

Important alloying elements like manganese, nickel, chromium, molybdenum, tungsten and vanadium have some effect or the other on the spark picture pattern of a steel. But it should be clearly understood that the phenomenon of sparking is fundamentally due to the presence of carbon. So, each of these alloying elements or in combination, modify the spark picture based upon the carbon content only.

Manganese : Manganese has not much effect on the spark picture of steel, in quantities less than 0.8 %. However, its presence can be noticed by the thin and light yellow carrier lines which spread out in a fan-like shape on leaving the grinding wheel (Fig. 7.21). However, if the carbon content is also high, star-bursts are noticed as in the case of a plain high carbon steel (Fig. 7.22).

Fig. 7.21 S.A.E. X-1315 (C 0.16, Mn 1.34).

(*Courtesy* : Iron Age, Radnor, Pennsylvania, U.S.A.)

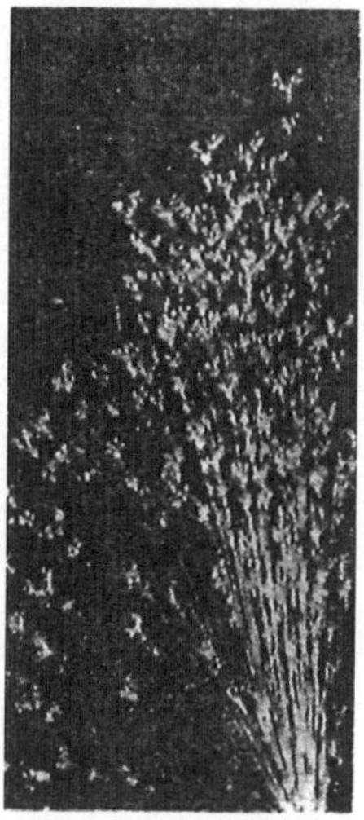

Fig. 7.22 S.A.E. 1350 (C 0.51, Mn 1.10).
(*Courtesy* : Iron Age, Radon, Pennsylvania, U.S.A.)

Nickel : Nickel swells the carrier lines, presumably trying to suppress the carbon bursts. The colour of the carrier line would be light orange, becoming darker as the percentage of nickel increases. The swelling of the carrier line would be so pronounced that it appears as rectangular blocks as the nickel content increases. Sprigs can also be noticed at the end of each carrier line, indicating the carbon content. They increase with the carbon content. These spark pictures are shown in Figs. 7.23, 7.24 and 7.25.

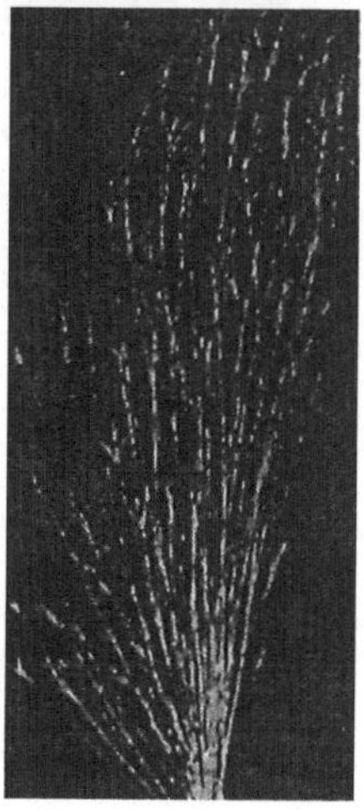

Fig. 7.23 S.A.E. 2115 (C 0.17, Mn 0.42, Ni 1.48).
(*Courtesy* : Iron Age, Radnor, Pennsylvania, U.S.A.)

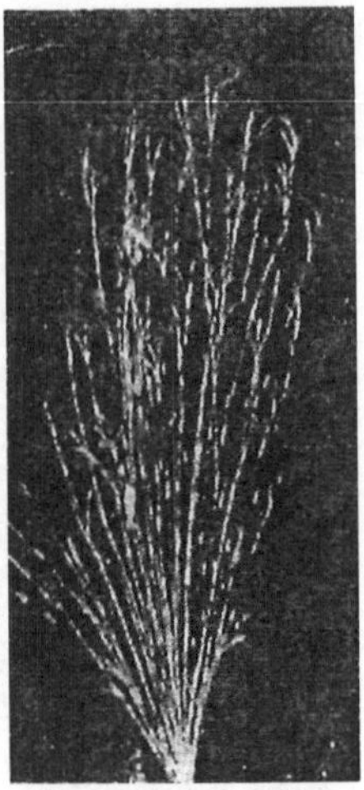

Fig. 7.24 S.A.E. 2320 (C 0.16, Mn 0.41, Ni 3.48).
(*Courtesy* : Iron Age, Radnor, Pennsylvania, U.S.A.)

Fig. 7.23 shows a low carbon 1.5% nickel steel, wherein the carrier line is thick, dark orange in colour, and exhibits sprigs at its end. Fig. 7.24 is the spark picture of a low carbon 3.5% nickel steel. The flame length and the number of carrier lines are more. Their colour would be still darker orange.

The pattern would be the same with high carbon varieties except that the carbon bursts will replace the sprigs at the ends of the carrier lines (Fig. 7.25).

Fig. 7.25 S.A.E. 2340 (C 0.42, Mn 0.69, Ni 3.47).
(*Courtesy* : Iron Age, Radnor, Pennsylvania, U.S.A.)

Chromium: When chromium is present in the steel, the carrier lines will be short, the colour of the flame is orange and a star shape is noticed at the end of each carrier line. This star shape should not be confused with the carbon bursts. The chromium stars are distinguishable on account of their greater brilliance and near-perfect star shapes. Further, the flame appears to flutter (Fig. 7.26).

Fig. 7.26 S.A.E. 3115 (C 0.17, Mn 0.53, Cr 0.68, Ni 1.21).
(*Courtesy* : Iron Age, Radnor, Pennsylvania, U.S.A.)

A low carbon and low chromium (0.6-0.9%) steel gives a spark picture just similar to that of a low carbon steel but with chromium stars (Fig. 7.26). As the carbon increases, the starbursts increase. Simultaneously, if chromium content is also increased, the spark picture consists of short carrier lines, dark orange in colour with flower-like stars and carbon bursts throughout (Fig. 7.27).

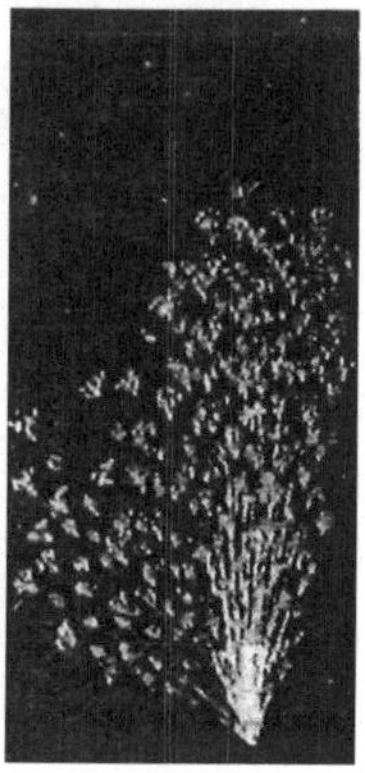

Fig. 7.27 S.A.E. 5210 (C 1.02, Mn 0.50, Cr 1.35).
(*Courtesy* : Iron Age, Radnor, Pennsylvania, U.S.A.)

When the steel contains both nickel and chromium, the characteristics of both the elements are noticed. As an example we can consider a steel with 3.5% nickel, 1.5% Cr, and 0.35% C (SAE 3335 variety) in which the spark stream is dull orange in colour, with the carrier lines bright and rectangular and chromium stars appearing at the ends (Fig. 7.28).

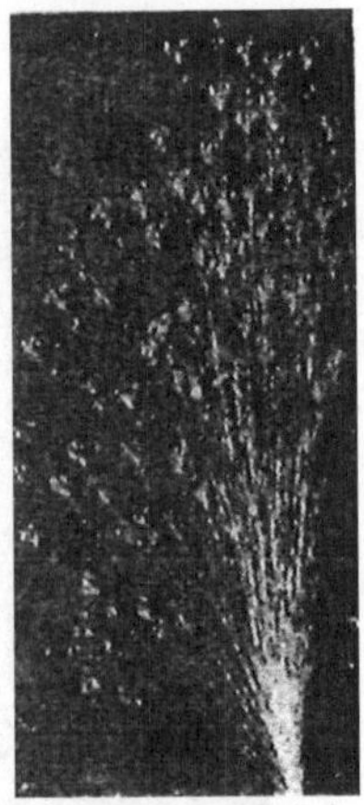

Fig. 7.28 S.A.E. 3335 (C 0.37, Mn 0.55, Ni 3.40, Cr 1.37).
(*Courtesy* : Iron Age, Radnor, Pennsylvania, U.S.A.)

Molybdenum: Molybdenum displays a carrier line shaped like a well-defined orange coloured spear-head at its end and detached from it. Fig. 7.29 illustrates the spark picture of a chrome-molybdenum steel (containing 0.5-0.8% Cr and 0.15-0.25% Mo) (SAE 4130 type). The detached spear-head and stars at the ends of the carrier lines can be noted.

Fig. 7.29 S.A.E. 4130 (C 0.32, Mn 0.69, Cr 0.75, Mo 0.22).
(*Courtesy* : Iron Age, Radnor, Pennsylvania, U.S.A.)

Vanadium : Vanadium also displays a carrier line similar to that of molybdenum except that the spear-head is at the end of the carrier line. A low carbon chrome vanadium steel exhibits a flame of dark orange colour which will be more fluttery than in the case of a straight chrome steel. With the increase of carbon percentage, carbon bursts start to appear in place of sprigs. The spark picture of a SAE 6120 (0.8 – 1.1% Cr and 0.18% V) steel is given in Fig. 7.30.

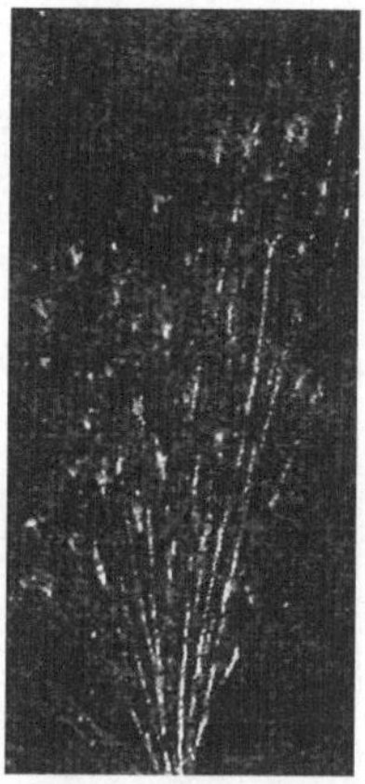

Fig. 7.30 S.A.E. 6120 (C 0.19, Mn 0.59, Cr 0.91, V 0.17).
(*Courtesy* : Iron Age, Radnor, Pennsylvania, U.S.A.)

Tungsten: The presence of tungsten suppresses the number of carrier lines. Even those few carrier lines would be short, tend to be straight and reddish in colour (Fig. 7.31).

Fig. 7.31 S.A.E. 7260 (C 0.55, Mn 0.28, Cr 0.95, W 1.97).
(*Courtesy* : Iron Age, Radnor, Pennsylvania, U.S.A.)

Silicon: Silicon is present in steels along with manganese. A typical silicon manganese steel SAE 9260 grade (0.57 % C, 0.81% Mn and 1.92% Si) exhibits a spark picture as in Fig. 7.32. Silicon suppresses the carbon bursts. The flame would be dark red and the carbon bursts tend to cling to the sides of the carrier lines. The carrier lines themselves are much shorter (in comparison to a plain carbon steel of the same carbon content) and usually end abruptly with a white flash (known as the 'dot').

Fig. 7.32 S.A.E. 9260 (C 0.57, Mn 0.81, Si 1.92).
(*Courtesy* : Iron Age, Radnor, Pennsylvania, U.S.A.)

Salient Features of Spark Testing

This testing method is for the classification of steels based on their composition with the aid of their spark picture. The test is very fast, convenient and economical. It is fairly accurate, however, subject to its own limitations. A great advantage of this test is that it is the only available test on hand when a material mix-up occurs. Another advantage of this test is that it can be applied to test steels in any stage of production, i.e., as ingots, billets or finished products. The most important factor in this is the human factor. The tester should be very keen and experienced.

References

IS : 1963 (pt. 1) – 1960, and (pt. 2) – 1962, Method for tensile testing of steel sheet and strip.

IS : 1894 – 1962, Method for tensile testing of steel tubes.

IS : 2655 – 1964, Method for tensile testing of copper and copper alloy tube.

IS : 2656 – 1964, Method for tensile testing of copper and copper alloy wire.

IS : 2657 – 1964, Method for tensile testing of aluminium and aluminium alloy tube

IS : 2658 – 1964, Method for tensile testing of aluminium and aluminium alloy wire.

IS : 1521 – 1960, Tensile testing of steel wire.

IS :1692 – 1960, Simple bend testing of steel sheet and strip less than 3 mm thick.

IS : 1403 – 1959, Reverse bend test for steel sheet and strip less than 3 mm thick.

IS : 1599 – 1960, Bend test for steel products other than sheet, strip and Wire.

IS : 1894 – 1962, Tensile testing of steel tubes.

IS : 2328 – 1963, Flattening test on steel tubes.

IS : 2329 – 1963, Bend test on steel tubes.

IS : 2330 – 1963, Flanging test on steel tubes.

IS : 2335 – 1963, Drift expending test on steel tubes.

IS : 1756 – 1961, Modification of Erichsen cupping test for steel sheet and strip.

IS : 3600 – 1966, Code of procedure for testing fusion welded joints and weld metal in steel.

PART - II

CHAPTER 8

Non-Destructive Testing

INTRODUCTION

Any component should be tested for its correctness. Correctness implies its satisfying the various specifications like dimensions, surface finish, mechanical requirements such as strength, hardness, ductility, impact, fatigue and creep properties, and soundness. Soundness of an object, if a cast one, means its being free from casting defects like blow holes, porosity, etc., and if a wrought one, not constituting seams, cracks and the like.

Testing for the mechanical properties will not always reveal the inherent unsoundness of a material. This is particularly true when a large latitude is provided for in the specification. Also, the determination of mechanical properties is not possible in all cases. In such instances arises the need to find out the soundness of an engineering component by other means. These methods of testing by which we do not spoil or damage the product and make it useless for service, are known as non-destructive testing techniques.

There is no, and cannot be, a rigid distinction between non-destructive testing and destructive testing. For the purpose of classification we put the various strength determination procedures under the destructive head and the others under the non-destructive category. Again, trouble comes up with hardness measurements. Brinell test on an auto-cylinder block is not destructive (unless it is performed on the bore surface after sectioning the block). The same holds good in the case of objects like a giant fly-wheel, a lathe body, etc. But, it is a thoroughly destructive test when performed on a bearing bush. However, it is the common practice to class Brinell, Rockwell and Vickers tests also, in a majority of cases, under the destructive group. Shore's Scleroscope is a non-destructive test, as the impression it leaves on the surface of the object is very small. Similarly, the various cupping tests and bend tests fall under the destructive category.

It can be seen further that the destructive tests yield us definite readings from which a direct understanding of the properties of the material is possible. For example, when a material gives a 0% elongation in a tensile test, we can readily call it brittle. In non-destructive testing, barring a few like visual, leakage and penetrant examination, which are mostly surface examinations, the test data should be properly interpreted. In many a testing process, the various physical properties of the metal or alloy are made use of. In the magnetic particle inspection, the magnetization of the object under an electric flux and the lines of flow are determined. From the pattern obtained, the defect, if any, its size, shape, and location are estimated. Thus, the interpretation of the test data is very important.

Non-destructive testing techniques have their own limitations. A magnetic test cannot be performed on a material which is non-magnetic. Thus, a comprehensive idea about the product under test, for what it is to be tested, what method of test will be most suited and its limitations, besides the advantages, should all be comprehended well in advance.

Above all, there should be a need for the test and the cost of testing should be justified. The relative saving which results, consequent to the testing, should be more than the investment made on the testing equipment and personnel.

It should be clearly decided which testing method is most suited for a particular purpose when more than one can be applied. This decision should be made taking into account the various advantages and disadvantages of the different methods.

To sum up, the decision regarding a non-destructive test should be made up of the following considerations:

1. **Necessity**

   Is non-destructive testing necessary or not?

2. **Material under test**

   Whether the material is magnetic or nonmagnetic, heavy or light, a plain material, a coated or plated one, its composition or constitution, and its service history should be explored. This helps, to a large extent, to decide what type of test is best suited for.

3. **Geometry**

   Size, shape and surface finish are the main considerations. These help us to use the right type of equipment, once the test to be performed is decided upon.

4. **Method/Methods of fabrication**

   The method of fabrication of the object is the consideration here. Whether it is cast, wrought, a powder metallurgical product, a welded joint, coated one, surface treated chemically or heat treated–largely determines its properties, types of defects that are likely to be present, and so on.

5. **Possible defects**

   A knowledge of the method of fabrication helps us to predict the likely defects that occur. The next step is to analyze them as to whether they are surface, sub-surface or internal defects.

6. **Reasons for the defects and their location**

   After all the above information is elicited and predicted, it is the final duty of the metallurgist to correctly imagine as to when and in what process or stage of manufacturing did the defect occur. Then it will be possible to exercise the needed care, control the defects, and avoid the recurrence of the defect. Thus alone, the fruits of testing are fully reaped.

It can be seen from the above that all the non-destructive testing procedures employ and base upon some property of the material. The response of the material to the physical, electrical or any other property in question is studied. Then the way this response gets modified in the presence of a defect is studied. Once the behaviour of the material is studied, it will be possible to predict the defect. Various non-destructive testing techniques can be classified into the following major types depending upon the material property being made use of, in the test.

NON-DESTRUCTIVE TESTS

The various types of non-destructive tests are enumerated below :

1. Visual examination,
2. Leakage testing,
3. Penetrant methods,
4. Magnetic particle inspection,
5. Sonic and ultrasonic (acoustic methods).
6. Radiography,
7. Thermal methods,
8. Electrical methods, and
9. Surface and thickness measurements.

CHAPTER 9

Visual Examination

This is the first type of test, performed knowingly or unknowingly. When we take a new product into our hands it is but natural to become curious about its appearance. We see it, or technically speaking–examine it.

Visual examination reveals the surface smoothness or roughness, and the surface defects. Defects like burnt-on sand, misruns, cavities which appear at the surface, etc., can be spotted by visual examination.

The experience of the examiner, his eye-sight, lighting conditions at the spot are the various factors which effect this testing method.

Artificial illumination is employed to avoid discrepancies in lighting. The surface is felt by hand, to feel the presence of a cleavage or crack, and smoothness.

Though the oldest method and involving no equipment, this method proves to be particularly useful in many instances. In fact, certain defects which cannot be properly spotted by very sophisticated methods of testing are located by simple visual examination, e.g., an uneven surface.

CHAPTER 10

Leakage Testing

Suppose there is a vessel which has a hole at its bottom. When we fill it with water, the water will start draining out from the bottom. We may not be able to even fill the vessel ! If the hole is a small one, we can fill the vessel, but the water leaks out. If the hole is too small to be easily perceptible, the water does not emerge as a thin jet but in the form of droplets. Thus, the magnitude of the outflow or leak is a direct indication of the size of the hole in the vessel.

In the last instance mentioned, suppose the vessel is pressurised, the water which hitherto was leaking in the form of small drops, can be seen to continuously ooze out. It can be further seen that if the pressure is increased, the outflow of water also increases. So, it can be said that it is possible to find out even a very small cavity in an object, if it is subjected to hydraulic pressure. This is the principle of testing many engineering items like auto-cylinders, pump bodies, etc.

The test is made on those components which, in service, are required to operate at and withstand high pressures.

The process consists of a simple device which gradually develops a high pressure. A pressure gauge is connected to read the impressed pressure. The casing under test is temporarily closed at all the openings excepting one. The open end is connected to the pressure pump. The pump is started and the required pressure is developed gradually. The maximum value of the pressure applied will be slightly more than the maximum operating pressure of the casting in service, and maintained at that value for a specified time.

During testing, it is keenly observed whether the casting 'weeps' or not. If it does not, it is deemed to have stood up to the test. Otherwise it fails and it should be rejected or salvaged.*

---

*Salvaging of castings is given elsewhere in the book.

CHAPTER 11

Penetrant Methods

Surface defects such as cracks, laps, laminations, seams, etc., can be tested for and discovered by the penetrant methods. When a liquid is applied to the surface of an object it penetrates into the defect, by the capillary action. The liquid penetration is influenced by its surface tension and viscosity. Besides, the nature of the discontinuity and its interior also affect the penetration.

The process consists of applying a thin oil penetrant to the surface under examination. Sufficient time is allowed to facilitate the seepage of the oil into the defects and the surface oil is removed by suitable means. Then a powder is dusted on to the surface which has a chemical or physical action with the oil. The oil in the crack absorbs the powder. When all the superfluous powder is again wiped out, a colouration can be seen near the crack due to the oil and the dusting powder. The contour of the colouration defines the shape of the crack, its intensity and depth. Thus, the process, in detail, consists of the following stages (Fig. 11.1):

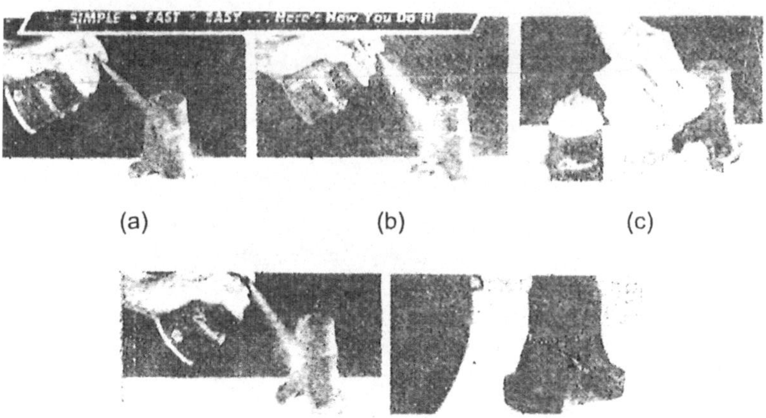

(a) (b) (c)

(d) (e)

Fig. 11.1 Sequence of operations in penetrant inspection using Zyglo. (*Courtesy* : Magnaflux Corporation, Chicago, Illonois, U.S.A.)

(i) Application of the penetrant liquid. The penetrants are either commercially patented ones* or made in the laboratory by mixing kerosene or diesel oil, cutting oil and oleic acid in certain proportions. Application is performed by dipping the part under test into the penetrant liquid. If the object is reasonably big or if only a particular surface is to be tested for, the penetrant liquid is applied either by brush, or by a pinch of cotton, or sprayed.

(ii) A few minutes time is allowed for the penetrant to seep into cracks.

(iii) The excess penetrant is wiped off from the surface and the surface is allowed to dry up.

(iv) A developer is dusted on to the surface, or if liquid, it is again applied, similarly as the penetrant liquid. Generally talc is dissolved in alcohol to make a developer.

(v) Some time is again allowed for the developer to be absorbed by the penetrant oil in the cracks.

(vi) The surface is again cleaned to remove the excess developer.

The defects, if any, can be observed with the naked eye or by means of a magnifying glass, depending on the type of developer employed (Fig. 11.2).

Fig. 11.2 Cracks in engineering components revealed by spotcheck penetrant. (*Courtesy* : Magnaflux Corporation, Chicago, Illinois, U.S.A.)

*A penetrant oil known by the commercial name 'ZYGLO' is made by the Magnaflux Corporation, U.S.A. After its use, ultra-violet light should be used for examining the crack. Other commercial penetrants like spot check, etc., can be used under white light only.

CHAPTER 12

Magnetic Methods

Principles of Magnetism and Magnetization

Bar Magnet

A magnet attracts magnetic materials like iron, steel, nickel, etc. It has two poles, one 'north' and the other 'south'. These two ends of the magnet are its most powerful regions and the force of attraction here is maximum.

Cut in a Magnet

A cut or slit in a magnet makes two more poles in the magnet. The iron particles are attracted near the cut also, as at the regular poles.

Horse Shoe Magnet

A bar magnet can be imagined to be bent in the shape of a horse shoe (U). The poles will still attract iron. The force of attraction will be more in this case because the two fields due to the north and south poles are concentrated adjacently.

Partially Closed Magnet

If the two ends of the bar magnet are bent still close to each other, the effect will be more pronounced. The attracting power of the two poles is practically concentrated at the same place.

Completely Closed Magnet

If the magnet is bent until the two poles meet and the ends are fused, it will no longer attract iron powder. In this case, there will be no polarity and the magnetic field will be confined in the circle inside the ring itself. This is called circular magnetization. However, if a cut is made in the ring, again polarity develops there and it attracts magnetic powder, just as a bar magnet with a cut.

Electricity and Magnetism

(i) When an electric current is passed through a wire, a magnetic field is produced around it. The direction of the magnetic field is at right angles to the direction of the current flow (Fig. 12.1).

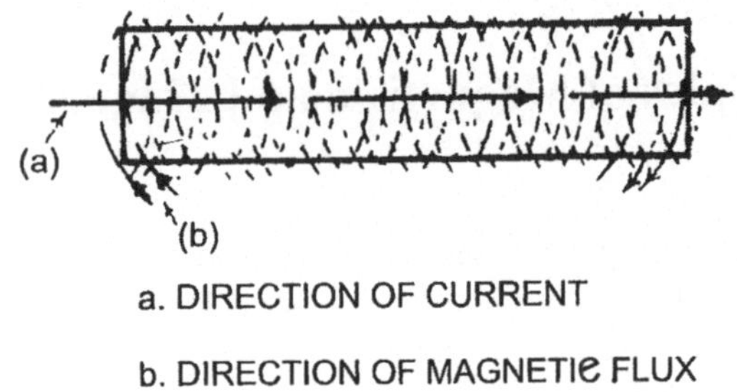

Fig. 12.1 Magnetic field at right angles to direction of current.

(ii) An electric current, when passing through a magnetic part, creates a magnetic field within the part. This is called circular magnetization.

(iii) When an electric field passes through a wire wound into the shape of a coil, it creates a magnetic field, in the direction of the axis of the coil. When a magnetic part is placed inside such a coil the magnetic field runs lengthwise in the part. This is known as longitudinal magnetization (Fig. 12.2).

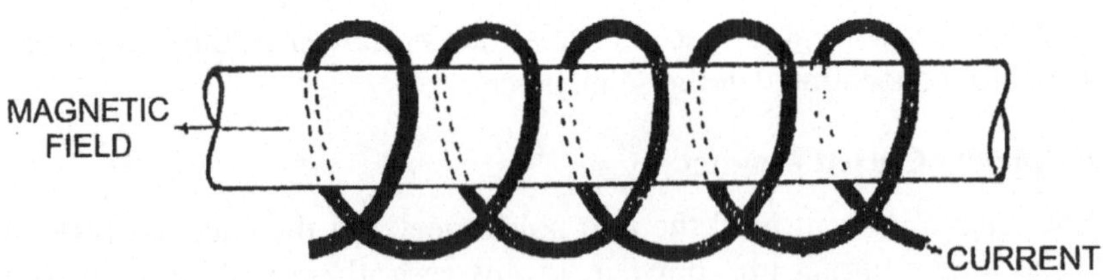

Fig. 12.2 Longitudinal magnetization.

Principles of Magnetic Particle Inspection

If an object which is made of a magnetic material is placed inside a coil carrying current, it will be magnetized, by the longitudinal magnetization effect. Any crack on its surface makes it behave as a bar magnet which has a cut. If magnet-powder is sprinkled over it or made into an emulsion and poured over this part, it is attracted more at the region of the crack. Thus, we can say that the defect is located at the regions where the magnetic particles are attracted and held most.

THE MAGNAFLUX MACHINE

The Magnaflux Machine manufactured by the Magnaflux Corporation, Chicago, U.S.A., is a patent and the one used most commonly. It consists of two heavy copper contact plates mounted on two heads, the left hand side one is fixed and the right one movable. The movable head is operated by a handle. The whole set up is mounted on a heavy body, much the same size of an ordinary lathe machine. A low voltage current, usually ranging from 4,000 A to 10,000 A, is passed through the part under test, the duration of the flow being controlled by a time relay switch.

The equipment has also a provision for a magnetizing coil to effect longitudinal magnetization of the part. The coil which is of about 50 cm diameter is mounted on the machine bed and moves on two sets of wheels. Below the bed, there is a tank into which collects the magnetic particle suspension applied to the job. Photographs of some models of magnaftux machines are shown in Figs. 12.5 and 12.6.

The Process

Longitudinal magnetization of the part is used by employing the magnetizing coil to detect any flaws which are transverse (Fig. 12.3), and circular magnetization (i.e., by passing the current through the part itself) is used to detect the cracks which run along the length of the part (Fig. 12.4).

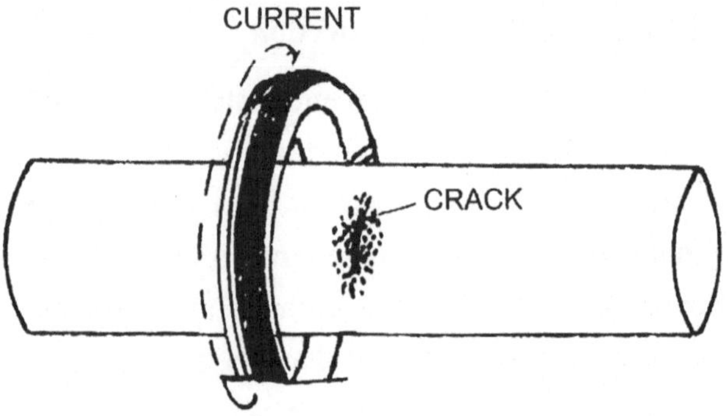

Fig. 12.3 Magnaflux indication of a transverse crack (in a cylindrical part).

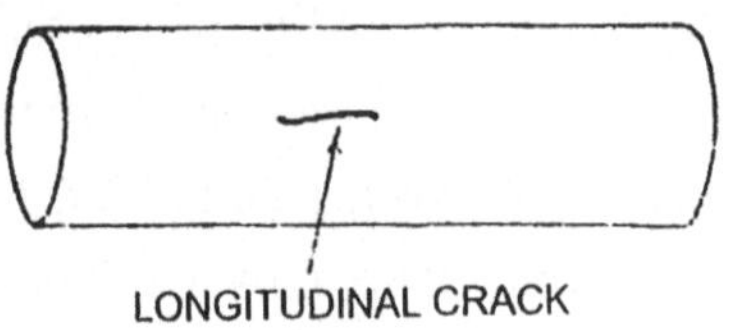

Fig. 12.4a.

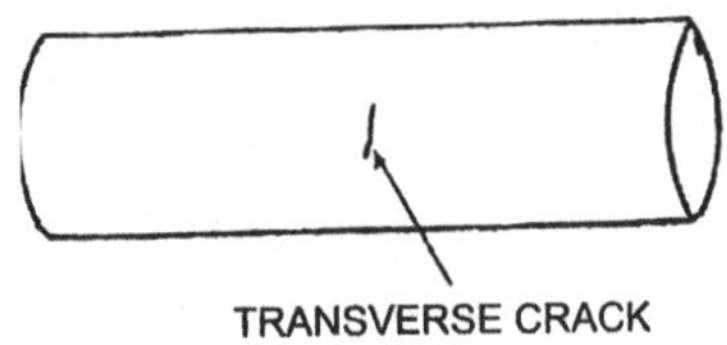

Fig. 12.4b

The Magnetic Bath

Magnetic iron oxide powder, fine enough to pass 100 mesh (ASTM) sieve, is the medium generally employed. The particles are given a suitable treatment so as to sligthly elongate them. Elongated magnetic particles offer a higher degree of polarization in a magnetic field. The powder is suspended in a thin liquid medium like kerosene.

'Magnaglo' powder and paste are the proprietary articles which can be used instead of the magnetic powder. The 'Magnaglo' paste is used in quantities varying from 0.5 to 1 .00 gm per litre of kerosene. The paste is weighed in a clean dish and is made thinner by adding a small quantity of kerosene to it. Later the thin paste is added to the remainder of the oil and thoroughly stirred, preferably in a mechanical agitator.

Method

There are two methods for the application of the magnetic bath:

1. Continuous method

The bath is applied to the part simultaneously with the current (Fig. 12.5). This is to facilitate the presence of the magnetic particles on the part when the current is being passed through it. The following are the various stages in the process :

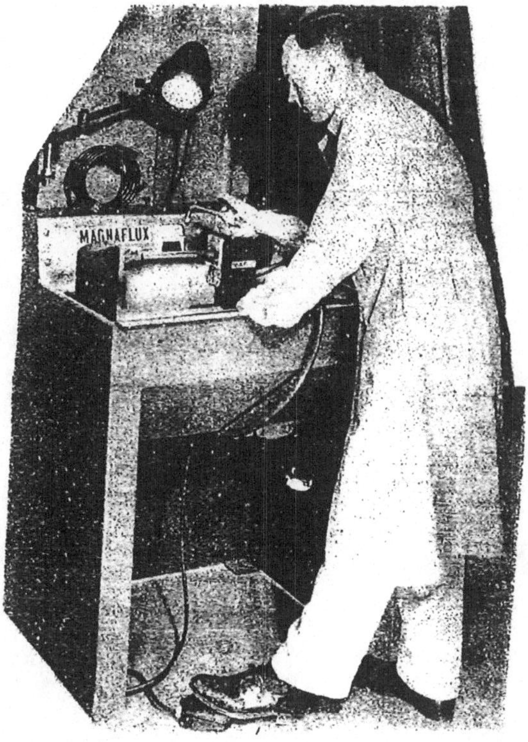

Fig. 12.5 Using Magnaflux machine. The inspector to ready to apply the magnetic particle emulsion onto the part magnetised.
(*Courtesy* : Magnaflux Corporation, Chicago, Illinois, U.S.A.)

(i) The part is held between the contact plates of the machine.

(ii) The bath is applied so as to uniformly wet the part.

(iii) The bath stream is stopped.

(iv) The current switch is closed and at the same time the bath is released onto the part.

(v) Current strengths of sufficiently higher magnitudes should be employed so as to reveal the defects in the part. At the same time, the current should not be so high as to build up flux lines at the regions of change of section.

2. Residual method

In this method, the bath is applied after the current has been shut off (Fig. 12.6). The magnetic particle suspension is applied when the part under test is a magnet by virtue of residual magnetism. The process in made up of the following stages:

(i) The current is passed into either the part or the coil so that the part under test is magnetized.

(ii) The current switch is opened.

(iii) The bath is applied to the part, preferably after it is removed from the machine.

Fig. 12.6 An auto-cylinder block being inspected using a magnaflux unit.
(*Courtesy* : Magnaflux Corporation, Chicago, Illinois, U.S.A.)

While the continuous method can be employed on any type of steels, the residual method is suitable only for testing magnetically 'hard' steels, viz., high carbon or alloy steels.

Demagnetization

A large proportion of the engineering components used are steels or other ferro-magnetic alloys. These are all magnetically hard, i.e., they do not lose their magnetism soon after the magnetizing force is removed. When stacked together they attract each other resulting in the associated difficulties. Thus, the parts should be demagnetized after testing, so that this residual magnetism is lost.

Residual magnetism is removed by reversing the polarity of the field in the part. This is possible when the testing adopted is of circular magnetization. The current direction is reversed and thereby the direction of the magnetic line of force. Thus, the earlier magnetic flux is cancelled out. If, on the other hand, the testing is done employing the magnetizing coil (i.e., longitudinal magnetization), or an alternating current, it is suggested to draw the part out of the coil gradually. The method leads to a gradual decrease in the flux density.

Application of the Method

The Magnaflux method is particularly suitable for the detection of surface defects such as cracks of all types (Fig. 12.7), seams, laps, laminations, overlaps, non-metallic inclusions, hot tears, etc. Some subsurface defects can also be detected.

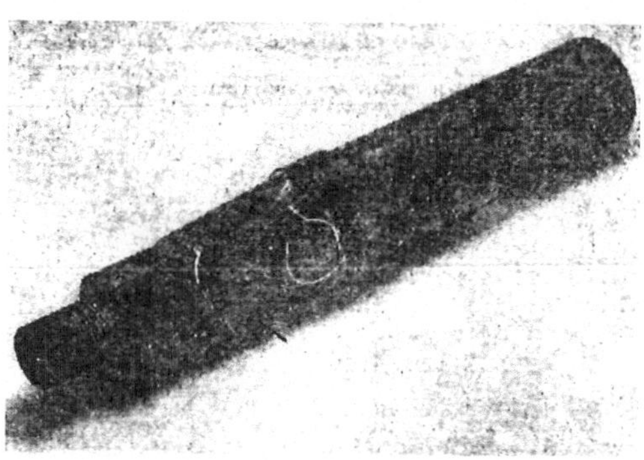

Fig. 12.7 Fatigue cracks as revealed by magnaflux method.
(*Courtesy* : Magnaflux Corporation, Chicago, Illinois, U.S.A.)

The tester should be very careful not to mis-interpret the result of the test. Products with complex surface treatments may give indications which will be grossly misleading (products which are selectively carburised). Further, a seam may be misinterpreted to be a wide sub-surface crack. These abberations should be scrupulously avoided, as said in the beginning itself. The effectiveness of non-destructive testing will be lost if misinterpretation is not avoided.

Salient Features of the Process

The magnaftux method is a relatively easy and simple technique, free from any restriction of size, shape, composition and heat-treatment of the product under test. The only limitation of the process is that it is exclusively for ferro-magnetic materials. Moreover, internal defects and defects which are deep-seated in the case of heavier sections are detected with difficulty.

Nowadays, small portable magnaflux units are available so that they can be carried to the work-spots (Fig. 12.8). These are especially useful in fabrication and other industries. The important adverse factor for this method of testing is the prohibitively high cost of the equipment.

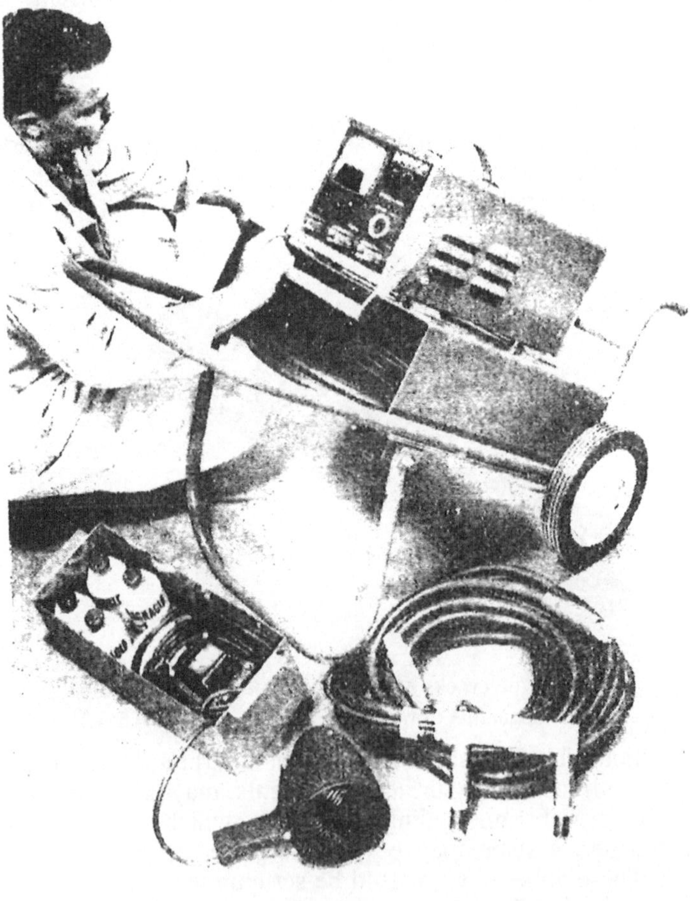

Fig. 12.8 Portable magnaflux unit.
(*Courtesy* : Magnaflux Corporation, Chicago, Illinois, U.S.A.)

CHAPTER 13

Acoustic Methods

Acoustic testing is performed using the sound energy. All the metals and alloys give a ringing sound when struck with a hard material. This is due to the vibrations produced in the metal.

Sound energy can be classified into two types, the audible and the inaudible. The waves are audible when their frequencies are lower than 20,000 c/s and above that frequency the sound energy is inaudible. The inaudible sound waves (having frequencies over 20,000 c/s) are called the ultrasonic or supersonic waves.

Accordingly, testing of a part by audible sound energy is called the sonic testing and by making use of the ultrasonic sound waves, supersonic testing.

SONIC TEST

The most commonly employed and oldest test of this category is the hammer test. When a solid metal or alloy of a homogeneous constitution is struck with a hammer, it emits a clear ringing sound. If the object is defective (having a crack or so) a dull sound is given.

This test is very suitable for objects of simple shapes. Modified sounds, likely to confuse the tester, may be emitted by articles of complex shapes.

ULTRASONIC TESTING

Supersonic or ultrasonic waves are produced by 'piezo-electric effect'. When crystals like quartz are placed in an electric field, a mechanical deformation is produced. When the current is alternating, the deformations or oscillations produced in the crystal will also be of the same frequency as the current. Thus, ultrasonic waves can be produced using high frequency currents.

Types of Ultrasonic waves

The ultrasonic waves are of two types :

1. Body waves, and 2. Surface waves.

Body Waves

The body waves are in turn of two types, longitudinal waves, and transverse waves.

The transmission of the ultrasonic energy in a material depends upon particle vibration. As the ultrasonic wave travels through the solid, the particles in the solid are displaced. When the longitudinal waves travel through a material medium, the direction of displacement of the particles of the medium will be the same as the direction of travel of the wave (Fig. 13.1). These waves are also known as compressional waves, dilational waves or irrotational waves.

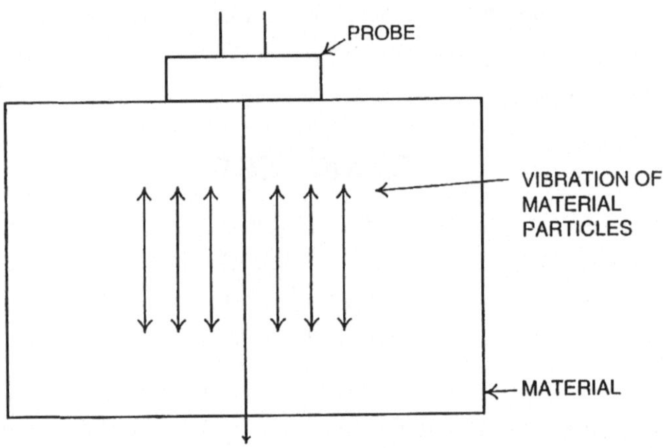

Fig. 13.1 Longitudinal wave.

When the transverse or shear waves propagate, the particles of the medium vibrate at right angles to the direction of the wave (Fig. 13.2). These are also known by the name distortional waves. An isotropic solid is capable of transmitting both the longitudinal and transverse waves. Since metals are isotropic in nature, both these types of waves can be used for the purpose of non-destructive testing.

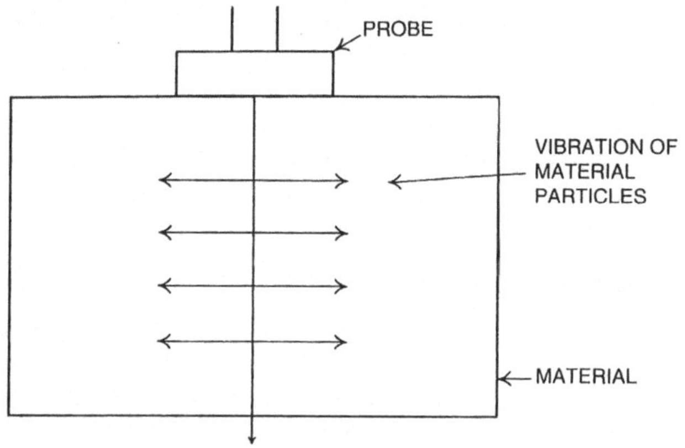

Fig. 13.2 Shear wave.

Surface waves

Ultrasonic waves of considerable amplitude can be propagated on the surface of a solid. These are called surface waves which are of three types–Rayleigh waves, Lamb waves, and Love waves.

Surface waves which propagate on a surface whose thickness is large in comparison to the wave lengths are called Rayleigh waves. Propagations of Rayleigh waves is best understood by comparison to the water waves in a lake. In this case, the motion of the particles of the medium is both longitudinal and transverse. However, the velocity of propagation of these waves is less than that of the body waves. (For metals, it is about 90% of that of the transverse wave velocity).

Surface waves which are propagated over the surface of a solid whose thickness is roughly the same as that of the wave length of the wave are called Lamb waves. A thin plate is capable of transmitting an infinite number of Lamb waves. The product of the thickness of the solid and the frequency of the waves determines the velocity or transmission of the wave through the solid.

Love waves travel on the surface of a solid with no vertical component. Their velocity of propagation is governed largely by their frequency; velocity decreases as the frequency increases.

FLAW DETECTION AND ULTRASONIC ENERGY

It is said above that the ultrasonic waves travel through solids. A very useful method of flaw detection by ultrasonic waves makes use of the phenomenon of the reflection of the ultrasonic waves by a surface. If the ultrasonic waves come across a surface at right angles to their path, they are reflected. Thus, a transverse crack or cavity sends back two reflections from its two surfaces.

The Ultrasonic Flaw Detector

The ultrasonic flaw detector makes use of the phenomenon of back reflection of the waves by surfaces. When the ultrasonic waves are made to pass through an object, they are immediately reflected by the surface at which they enter. Then, after they reach the other end of the solid, they are again reflected back.

If the solid has a crack or cavity inside, the two surfaces of the cavity also hinder the wave flow and reflect them (Fig. 13.3).

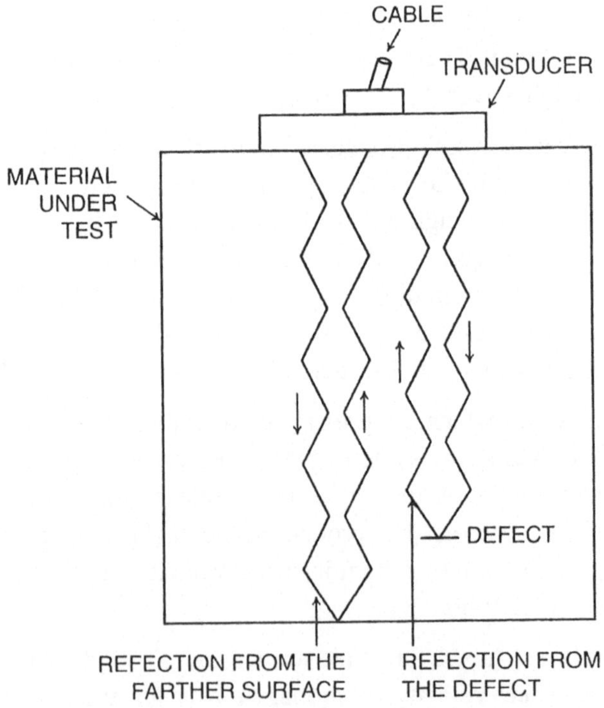

Fig. 13.3 Reflection of ultrasonic wave from a defect.

One crystal probe is sufficient to transmit into, and receive from the solid, the ultrasonic waves. The time of application of the current and its frequency are suitably controlled to send a predetermined number of ultrasonic waves into the object under test. Thus, if a 5,000,000 cycle frequency current is used, ultrasonic waves of the same frequency are generated in the piezoelectric crystal. When applied for a millionth of a second, five such ultrasonic waves are sent into the surface of the object. When the current is put off, the crystal ceases to produce and transmit ultrasonic waves and starts to function as a receiver. The reflected ultrasonic waves generate vibrations in the crystal, thus producing electric impulses. These are fed into a cathode ray oscilloscope. The crystal probe or transducer can be selected from a large variety, depending upon the purpose. Different types of transducers are shown in Fig. 13.4.

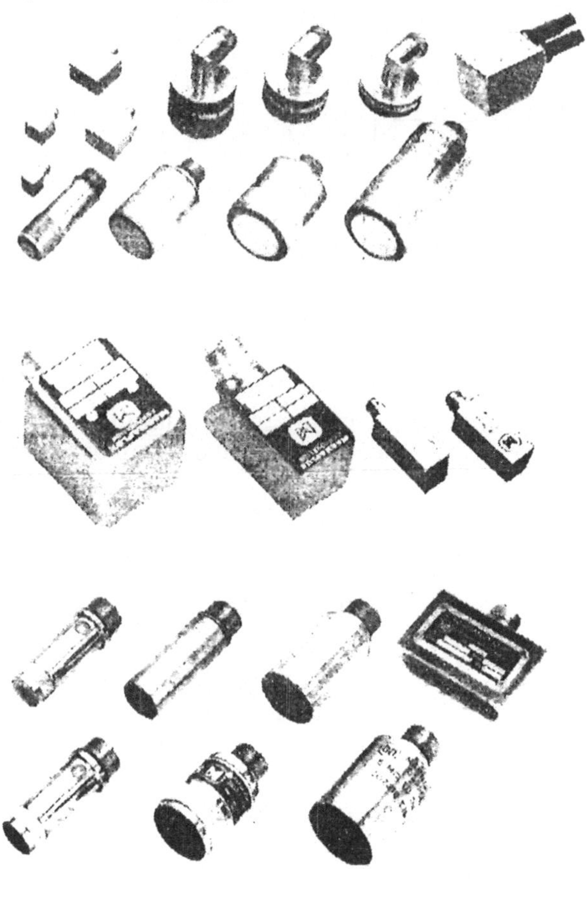

Fig. 13.4 Ultrasonic testing probes.
(*Courtesy* : Magnaflux Corporation, Chicago, Illinois, U.S.A.)

The oscilloscope is timed suitably depending upon the number of repetitions of the application and receiving of the ultrasonic waves into the object. The oscilloscope then sweeps to the right the same number of times per second. The initial vibration, on being reflected from the entering surfaces, makes a sharp peak at the extreme left side of the oscilloscope screen. The waves, reflected from the other (farther) surface of the object, produce another sharp peak at the extreme right of the oscilloscope screen (Fig. 13.5a).

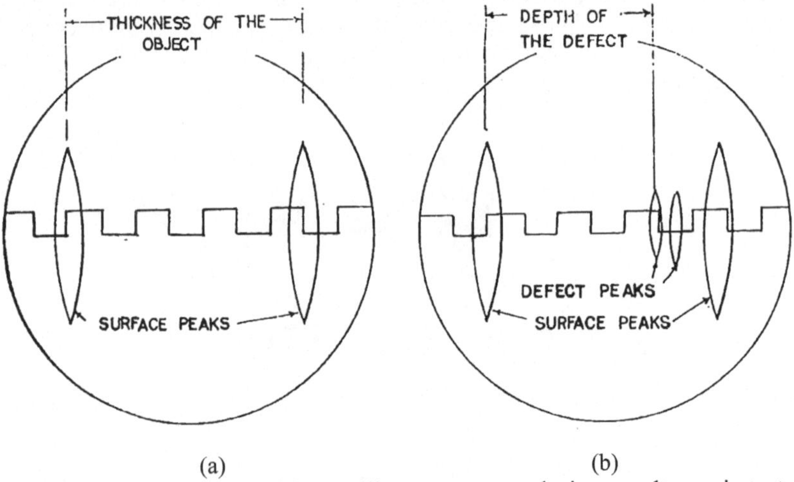

Fig. 13.5 Appearance of the oscilloscope screen during an ultrasonic test.

If the object has a defect falling in the path of the supersonic beam, two additional peaks are noticeable on the oscilloscope screen besides the ones on both the sides (Fig. 13.5b). This may also be recorded by a suitable ancillary mechanism (Fig. 13.6).

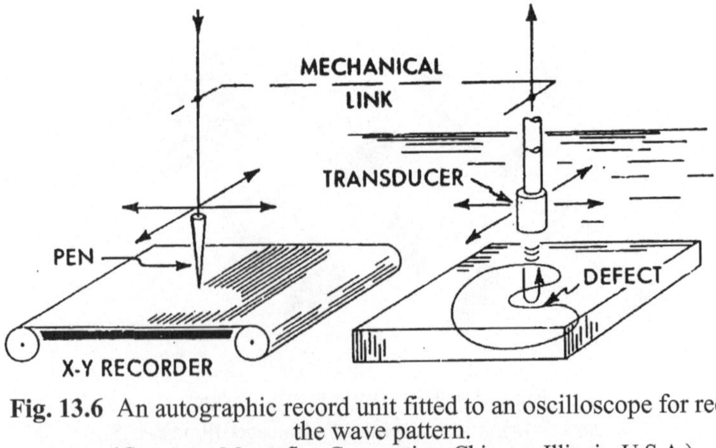

Fig. 13.6 An autographic record unit fitted to an oscilloscope for recording the wave pattern.
(*Courtesy* : Magnaflux Corporation, Chicago, Illinois, U.S.A.)

The location of the defect can be accurately made by noting the relative distances of the defect peaks from the other two extreme peaks.

The ultrasonic beam emanating from the crystal is very narrow. It is thus necessary to cover every bit of the surface of the object by a progressive movement of the probe. Further, it is also necessary to repeat testing process on another surface at right angles to the face once tested, so that the defects which might have been oriented longitudinally to the path of the ultrasonic waves in the first instance, are also revealed.

Precautions

Some precautions must be taken to ensure efficacy:

1. The surfaces of the object under test should be thoroughly cleaned.
2. An oil coating should be given to the surface being probed, to facilitate a thorough contact between the surface being probed and the probe.
3. Correct adjustment of the oscilloscope, depending upon the number of repetitions of the application and receiving the ultrasonic waves, is necessary. Otherwise, an exact location of the depth defect becomes difficult.

Salient features of the process

The process has no limitations in application. It is suitable to test any solid material, of isotropic nature. All types of defects are traceable by this method.

In spite of its high investment cost, the ultrasonic detector is very useful in a manufacturing organization. A crank shaft or cam shaft forging can be tested and scrapped if defective, without sending it to the machine shop. The machine shop operations in such cases constitute a giant share of the cost of processing and when eliminated in unnecessary instances, are a great saving in themselves.

A portable ultrasonic unit is shown in Fig. 13.7. Besides testing the products in the process of manufacture, the ultrasonic tester can also be used to check for any damage in service. Fig. 13.8 illustrates a maintenance engineer checking a tanker in a chemical plant for corrosion with the help of a portable ultrasonic unit.

Misinterpretations

Sometimes, spurious indications will be observed while inspecting metal parts. A sound part being rejected as a defective one is a colossal waste and every effort should be made to avoid this. Spurious indications are likely when the material being probed has a curved surface on the farther side. This may lead to the recording of an additional peak due to the double reflected transverse wave [see Fig. 13.9 (in the middle)] and the tester is likely to mistake it for a small crack. Such a peak can be found to decrease, in size if a finger is placed on the curved surface. It can also be construed as a spurious defect if a large number of similarly shaped products are tested and the indications are that all of them are unsound.

Another type of spurious indication will occur when the part under test has holes drilled in it which are filled with a liquid. Two reflections will be obtained from the metal liquid interface and the liquid metal interface. As the velocity of sound in the liquid is much less than in the metal, the two peaks will be distantly spaced. So, the second peak corresponding to liquid metal interface may be mistaken to be a defect.

Fig. 13.7 A portable ultrasonic unit in use in routine maintenance.

(*Courtesy* : Magnaflux Corporation, Chicago, Illinois, U.S.A.)

Fig. 13.8 A maintenance engineer checking a chemical plant for corrosion using a portable ultrasonic testing unit.
(*Courtesy* : Magnaflux Corporation, Chicago, Illinois, U.S.A.)

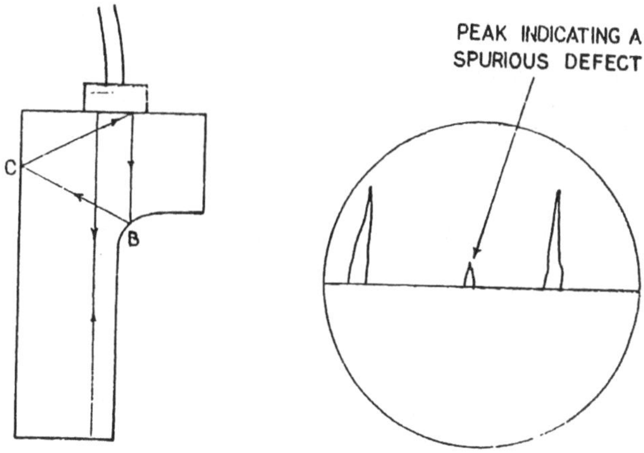

Fig. 13.9 Spurious indication due to the transverse wave generated at B and reflected at B and C.

Yet another type of spurious indication may occur when the object tested is long and the pulse repetition frequency is high. These indications will be haphazard or not clear. They can be easily identified as spurious.

CHAPTER 14

Radiography

X-rays and γ-rays occupy the invisible part of the electromagnetic spectrum. The electromagnetic spectrum can be roughly divided on the basis of wavelength into the following categories:

Gamma rays	:	0.01 - 0.005	A.U.
X-rays	:	5 - 0.008	A.U.
Ultraviolet rays	:	10 - 4,000	A.U.
Visible rays	:	4,000 - 7,500	A.U.
Infra-red rays	:	7,500 - 4×10^6	A.U.
Radio waves	:	10^6 - 10^8	A.U.

The wavelengths, being very small, are expressed in what are called the Angstrom Units; one Angstrom is equal to 10^{-10} metre or 10^{-8} cm.

Radiography is a process of non-destructive testing which utilizes X-rays or γ-rays. X-rays and γ-rays possess the following properties:

1. They are invisible.
2. They can penetrate matter.
3. They are differentially absorbed by matter.
4. They travel in straight lines.
5. They effect photographic papers and emulsions.
6. They ionize the gases while passing through them.
7. Electric or magnetic fields have no effect on them.
8. They travel with a velocity of 8×10^{10} cm/sec (186,000 miles/sec.)
9. They are capable of liberating photo electrons.
10. They can cause fluorescence in certain substances.

The properties with numbers 2, 3, 4, 5 and sometimes 10, to a large extent, are taken advantage of in radiography.

X-RAY AND γ-RAY RADIOGRAPHY

Though γ-rays have a greater penetrating power than X-rays and scatter less, they are not preferred in radiographic work since a greater time of exposure is needed. Again, the definition of the image is not satisfactory in the case of γ-rays compared to the X-rays. Portable X-ray equipment developed of late, has made the use of X-rays all the more easy and convenient. In respect of cost, however, the cost of a 300 K. V. X-ray unit and that of a γ-ray capsule with a 200 mg radium salt, works out to be the same. For small sections and those of uniform thickness, X-rays are satisfactory. For castings of irregular shape and varying cross sections, γ-rays are preferred due to less fogging.

The most universal application of radiography is in the testing of welds and castings. In the former case, there is generally no question of variation of section. In the case of castings also, this problem does not generally arise. Thus, X-ray radiography has become very common nowadays.

Production of X-rays

X-rays are produced when high speed electrons, called cathode rays, strike a metal object, called the target. When this happens, a transfer of energy occurs and X-rays are produced. Thus, for the production of X-rays, we need:

(a) an electron source,

(b) some means to accelerate the electrons, and

(c) a metal target.

The whole set-up is an X-ray tube which consists of an anti-cathode (for the production of cathode rays) and a target (Fig. 14.1). A difference of potential is maintained between the two so as to accelerate the electrons. This difference of potential is known as the tube voltage.

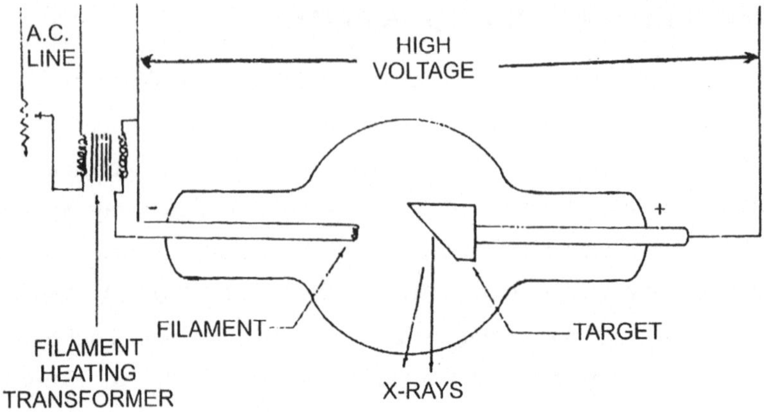

Fig. 14.1 X-ray production.

Since we have an electron flow in the X-ray tube and an electron flow is electric current, the magnitude of the moving electrons is determined by the tube current.

Flow of electrons is a process involving high velocity. Any air in the tube will offer resistance to the electron motion and defeat the purpose of accelerating the electrons. This leads to the scattering of the X-rays and focusing is lost. That is why the electron tubes are evacuated.

Only about 3% of the total electrons flowing are converted into X-rays. The remainder of the electron energy is wasted as heat. As a result, the target gets unduly heated up and, so, is water cooled.

The energy of the X-rays or γ-rays is expressed in terms of electron volts 'eV', Kiloelectron volts 'KeV', or million electron volts, 'MeV'. An electron volt is the kinetic energy acquired by an electron when it is accelerated in an electric field produced under a tube voltage of one volt. The different units are related as follows :

$$1 \text{ eV} = 1.6 \times 10^{-12} \text{ Ergs.}$$

$$1 \text{ KeV} = 1{,}000 \text{ eV.}$$

$$1 \text{ MeV} = 10^{6} \text{ eV.}$$

The energy of the accelerated electron is proportional to the square of its velocity. Thus, the velocity 'v' acquired by an electron under a tube voltage 'V' can be calculated as follows :

$$\frac{1}{2} Mv^2 = 1.6 \times 10^{-12} V$$

where M = mass of the electron (9.1×10^{-28} gms.)

v = its velocity

and V = tube voltage.

X-ray Tube

The two terminals are connected to a high voltage source; a milliammeter '*A*' is included in the circuit. A separate low voltage source supplies current to the filament. The current can be varied by manipulating the variable resistance '*R*' and its value is indicated by a milliammeter. There is a spark gap of about 3 to 10 inches with currents of 0 to 50 milliamperes and potentials of 100,000 to 300,000 volts.

The voltage applied to the tube and the current passing through the filament can be varied independently.

When the filament gets incandescent due to the low voltage current, electrons are emitted. In the prevailing high voltage field they get accelerated and strike the target to produce the X-rays.

A portable industrial X-ray radiographic unit is shown in Fig. 14.2a and its actual usage in the inspection of a large pressure vessel is illustrated in Fig. l4.2b.

Fig. 14.2a Portable industrial X-ray radiographic unit.
(*Courtesy* : Balteau, Belgium.)

Fig. 14.2b Inspection of large pressure vessel with portable industrial X-ray radiographic unit.

(*Courtesy* : Balteau, Belgium.)

The Radiograph

It is said above that X-rays and γ-rays penetrate matter and in that process are differentially absorbed. The absorption of the X-rays or γ-rays is described by the equation:

$$I = I_o \exp.^{(-\mu X)}$$

where I_o = intensity of the incident radiation,

I = intensity of the transmitted radiation,

X = thickness of the material through which the radiation penetrates, and

μ = linear absorption coefficient of the material.

Suppose an X or γ radiation of intensity I is penetrating through a rectangular metallic section of thickness D (Fig. 14.3). Further suppose that the section contains a rectangular cavity of width 'd'. Consider three rays A, B and C. The rays A and C are so incident that they do not pass through the cavity, for the rays travel in straight lines. The ray B travels through the cavity. The absorption of the energy while the ray is traveling through air or vacuum is practically nil in comparison to that while travelling through the metal. Thus, the ray *B*, after it emerges out, will have travelled, through a metal thickness of $(D - d)$ only whereas the rays *A* and *C* have traversed the full distance *D*. If the intensities of the transmitted rays be represented by I_D and I_d we can easily conclude that I_d will be more than I_D.

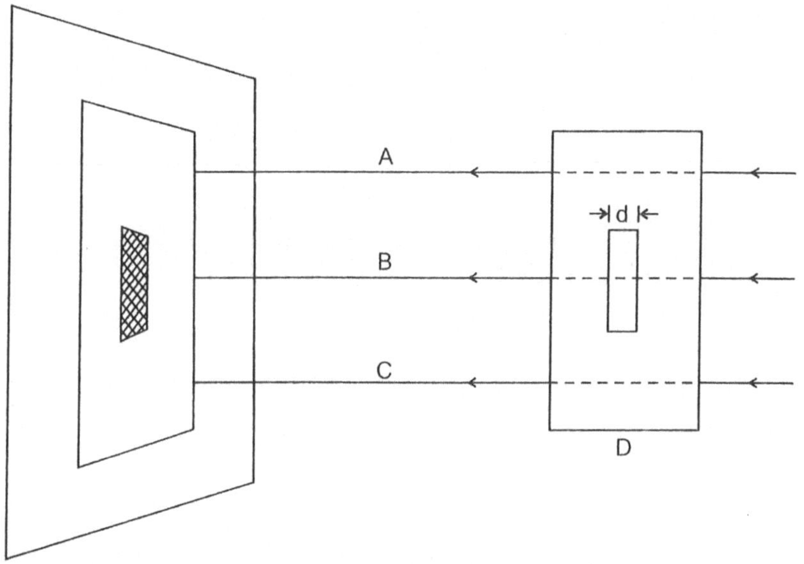

Fig. 14.3 Absorption of X-rays.

Under such conditions, if we place a photographic plate in the path of the transmitted radiation, the region of the cavity will be affected more than the remainder of the portion. This is the principle behind taking a radiograph.

Optical Factors which Effect the Radiograph

The radiograph is a shadow picture. It is affected by the relative positions of the object under test and the photographic plate and by the size of the focal spot of the X-ray tube.

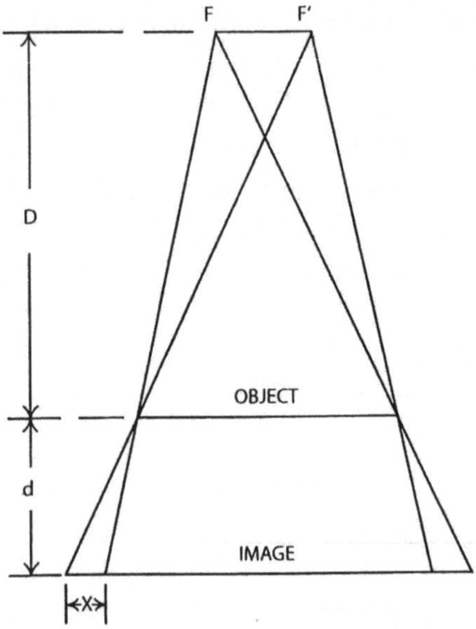

Fig. 14.4 Geometrical unsharpness

From Fig. 14.4, we can see that an unsharpness amounting to x is produced. The degree of unsharpness can be calculated from the similar triangles relationship thus:

$$\frac{FF'}{D} = \frac{x}{d}, \text{ i.e., } x = \frac{(FF')d}{D}$$

where x = unsharpness

FF' = size of the focal spot

D = distance between the object and the X-ray source, and

d = distance between the object and the X-ray film.

From the above relationship, we can say that the unsharpness will be the least when :

(i) the focal spot is small,

(ii) the distance between the object and film is minimum, and

(iii) the distance between the focal spot and the object is more.

Thus, we conclude that D/d ratio should be more in order to get a clear 'definition' in the radiograph. Definition is the degree of detail in the picture and describes the sharpness of the image. Any deviation of the image from the true shape of the object is called the 'distortion'.

X-Ray Films

An X-ray film consists of a thin, uniformly coated silver bromide gelatin emulsion on a transparent cellulose acetate base. The coating thickness does not exceed 1/250th of a millimetre.

The X-ray films can be classified into two major types :

1. The fine grain films, and
2. The coarse grain films.

In the fine grain films, the coating is made up of very fine crystals of silver bromide. These films need a longer time of exposure. Similarly, the coarse grain films are made up of relatively coarse crystals of silver bromide in their coating. With these films, a lesser time of exposure is, however, sufficient. Thus, depending on the exposure timings, these fine grained and coarse grained films are also called the slower and faster films.

Besides, there are also two other types of X-ray films. One is used with intensifying screens and the other types are meant for ordinary use.

Filters and Screens

A thin sheet of metal is placed between the source of radiation and the object. This is known as the 'filter'. Filter is used to absorb the longer wavelength. It is the property of the X-rays that as the wavelength increases, the degree of scattering too increases. The filter is thus useful to decrease, if not eliminate, the scattering.

If the same metal sheet is placed between the object and the photographic film, it becomes a screen. These screens are useful in making a more pronounced effect of the X-rays on the photographic plate. The X-rays which are incident on a photographic plate pass through it straightaway, rather than producing the required photochemical effect. This happens to about 99 percent of the X-rays. On the other hand, if a metal sheet is placed near the photographic plate, electrons will be produced when the X-rays fall on the plate. These electrons cause a photo-chemical effect on the plate. The metal generally used is lead. These secondary X-rays or electrons are more efficient in producing the photo-chemical change on the film than the primary electromagnetic radiation. With these screens a low energy radiation is to be used. For the best results, these screens should be placed in contact with the photographic film or else the electrons coming out of the screen may cause a blurred image.

These lead intensifying screens are generally used in pairs. The film is placed in between them. The screen which is placed towards the object is called the front screen and the other back screen. The back screen is made of a thicker sheet than the front one. This helps to totally stop the radiation from going out of the casette. The front screens are usually 0.02 mm thick while the rear ones range from 0.02 mm to 0.06 mm in thickness.

The other type of intensifying screens are called the fluorescent variety. Calcium tungstate is a chemical which exhibits fluorescence. It responds to the X-rays and fluoresces white light. The intensity of the bluish white fluorescence depends upon the intensity of the X-rays incident upon it. The screens made out of this material have two advantages–thicker sections can be radiographed with X-ray units of a moderate capacity and reduced exposure timings.

The use of these fluorescent intensifying screens is, however, limited owing to the poorer definition obtained in the radiograph. This is the result of the spreading of the visible fluorescent light emitted by the screen.

Taking a Good Radiograph

It is seen above that the X-rays are differently absorbed by different materials. Further, they are differently absorbed by the different thicknesses of the same material. We get an idea of the defect, if any, in the material, if its (object's) thickness is more or less uniform. In the case of a material which has a section of varying thickness, the intensity of the transmitted radiation will be more from the regions of minimum thickness. In such a case, we are likely to mistake those regions to be defective. Suitable techniques are to be adopted depending upon the shape of the object under examination. An object of circular section may be enclosed in pads containing hemispherical surfaces and made out of the same material as shown in Fig. 14.5. Thus, the X-rays will pass through a uniform thickness.

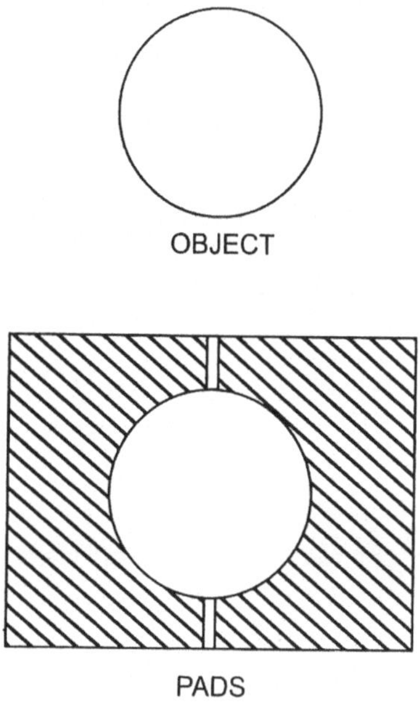

Fig. 14.5 Pads for use with irregular objects.

Irregular objects which cannot be modified to appear uniform in thickness as above, are, however, immersed in a lead salt solution. This helps, to a large extent, in obtaining a sufficiently uniform intensity. If this method is also not practicable, it is suggested that γ-ray radiography be adopted.

Sensitivity of a Radiograph

We take a radiograph of a metal to investigate any defect in it. The radiograph may be capable of revealing a small defect or only large defects, the words small or large being used with respect to the thickness of the object. When the radiograph can picture relatively minute defects we say that the "sensitivity" is high, and vice versa. Sensitivity varies depending up on the way in which the image contrast and sharpness are regulated. Thus, it is an important criterion in radiography and an idea of the sensitivity is necessary to reproduce the different forms of the defects in precisely controlled sizes in different materials and sizes. Unfortunately, however, it is not possible to directly measure the sensitivity.

A small device that helps to obtain a variation in thickness of the material artificially is the penetrameter, shown in Fig. 14.6. This is made of the same material as is under examination and placed on the top surface of the object, right above the area being probed. The resultant radiograph thus contains the image of the penetrameter also. The clarity of the penetrameter image will be found to diminish as the steps become thinner. The number of clearly visible steps is noted and the thickness of this portion is computed. The radiographic sensitivity of the material is the percentage of the above computed thickness to the thickness of the object under test.

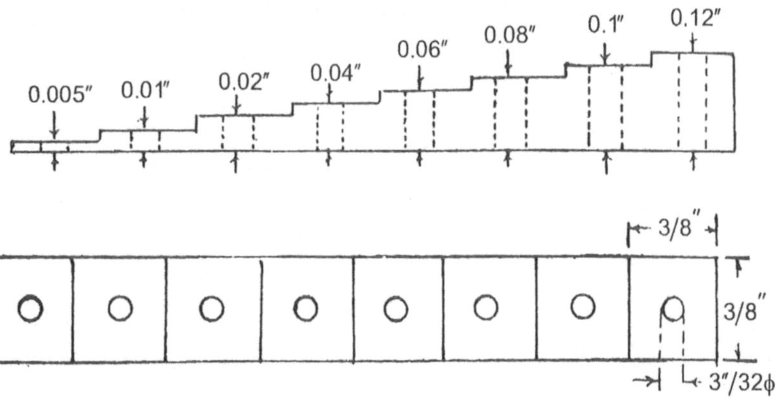

Fig. 14.6 Penetrameter.

However, no penetrameter can reproduce the type of image cast in a radiograph by defects such as porosity, segregation or sand inclusions. That is why we cannot take the penetrameter sensitivity as a measure of the defect sensitivity, but only a guide to assess the quality of the radiograph.

Production of γ-Rays

Radioactive metals like radium and their salts decay with time. During this process they emit γ-radiation.

The most commonly used material in the γ-ray units is radium sulphate. This is sealed in a silver capsule which, in turn, is enclosed in a duralumin container. This helps in easy handling and protects against loss. Commercial γ-ray radiographic units contain radium sulphate varying from 100 to 300 milligrams. The intensity of the radiation obtained is directly proportional to the quantity of the salt used.

The apparatus for the γ-ray radiography is very simple. The radioactive container is rigidly fixed in a stand to face the object under test. Suitable films are placed behind the object.

Interpretation of the Radiograph

The most important aspect of radiography is the interpretation of the radiograph. The radiograph gives a negative picture of the object. The dark regions in the negative correspond to those in the object whose thickness is less compared to the lighter ones. That is why less energy is absorbed by them and the radiation traversing through them had more intensity to affect the photographic film. These dark regions are also caused by the differences in the cross-section and should not be mistaken to be defects. Again, when the object under examination is relatively thin and possesses a rough surface, it may lead to lighter and darker areas in the radiograph. The surface roughness should be taken into cognizance and not mistaken to be any internal irregularities. Some common defects in castings and their appearance in the radiograph negative are as follows:

Sl. No.	Defect	Appearance
1.	Gas cavities, blow holes	Well-defined circular dark areas.
2.	Shrinkage porosity	Fibrous irregular dark region at the center.
3.	Cracks	Dark areas of varying width.
4.	Sand inclusions	Dark or grey spots of uneven texture with no clear boundaries.
5.	Inclusions in steel	Dark areas of definite outline.

Regarding the testing of welds, which is an important application of radiography, the common defects that arise are porosity, slag inclusions, cracks and incomplete fusion. It should be noted that all these defects are the regions which have a much lower density than steel. Hence they produce regions of darkness in the radiograph as those in castings do. Incomplete fusion, however, gives a dark line running parallel to the direction of the weld joint. Some photographs depicting weld-defects in steel are illustrated in Figs. 14.7, 14.8 and 14.9, while the photograph (Fig. 14.10) illustrates a shrinkage cavity in a casting.

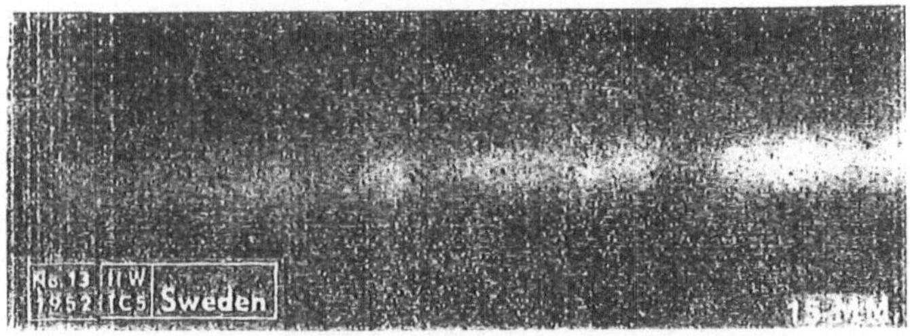

Fig. 14.7 Porosity. (Radiograph taken from the "Collection of Reference Radiographs of Welds in Steel," compiled by Commission V, "Testing, Measurement and Control of Welds" of the International Institute of Welding, London, U.K.)

Fig. 14.8 Slag inclusions in a steel weld. (Radiograph taken from the "Collection of Reference Radiograph of Welds in Steel," compiled by Commission V, "Testing Measurement and Control of Welds" of the International Institute of Welding, London, U.K.)

Fig. 14.9 Lack of fusion or improper weld penetration. (Radiograph taken from the "Collection of Reference Radiograph of Welds in Steel," compiled by Commission V, "Testing Measurement and Control of Welds" of the International Institute of Welding, London, U.K.)

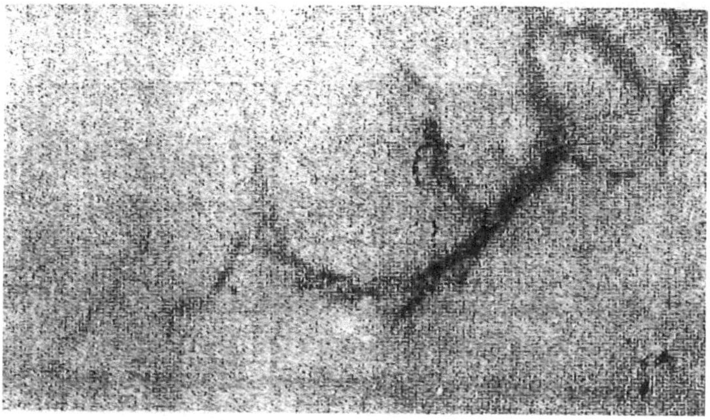

Fig. 14.10 Radiograph of a shrinkage cavity.
(*Courtesy* : Indian Copper Information Centre, Calcutta.)

Safety Precautions

X-rays and γ-rays are very dangerous from the point of view of human health. They kill the white blood corpuscles. Exposure to these radiations causes diseases like anaemia, leukaemia and low blood pressure and in extreme cases may cause severe skin rashes which may be fatal. The radiographers should undergo occasional blood count checks-up.

The X-ray radiation should be shielded heavily. A 5 mm thick lead shield should be employed both in front of and behind the photographic film.

Radium and its salts should be very carefully handled. When not in use, the radioactive container should be placed in a lead box which has atleast 15 cm thick walls. When the radioactive container is removed from the box, the operator should be at least three to four metres away from the container, unless he himself is heavily shielded. Even while fixing the container for taking the radiograph, it should not be handled even with gloved hands. Suitable sticks or wire devices should be employed. It is advisable that the radiographer should not remain even at the suggested distance for longer times than is absolutely necessary.

CHAPTER 15

Thermal Tests

Metals and alloys are conductors of heat. When a metallic object is heated at one end, the heat is conducted from this portion to the other portions. The tendency is to attain a uniform heat distribution in the body. After some time, the body attains a uniform temperature throughout. Thus if the temperature at the end where the heat is applied is 'T_1' and at the farthest end is 'T_2' the heat flow takes place and continues as long as T_1 is greater than T_2 (Fig. 15.1).

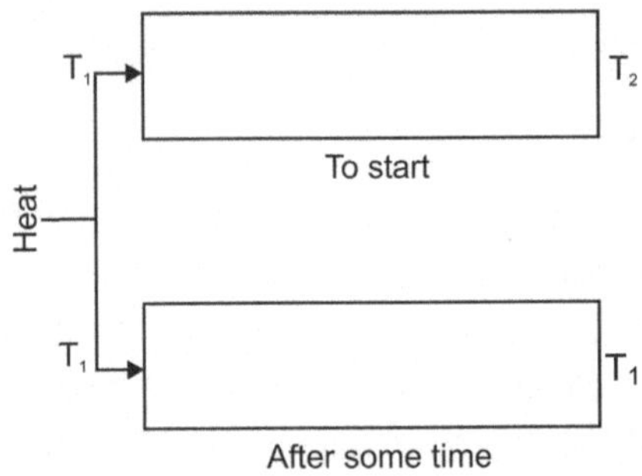

Fig. 15.1 Tendency of body to attain uniform temperature.

If there is a crack or discontinuity in the metal, the heat travel will not be facilitated. The region of discontinuity offers resistance to the heat flow.

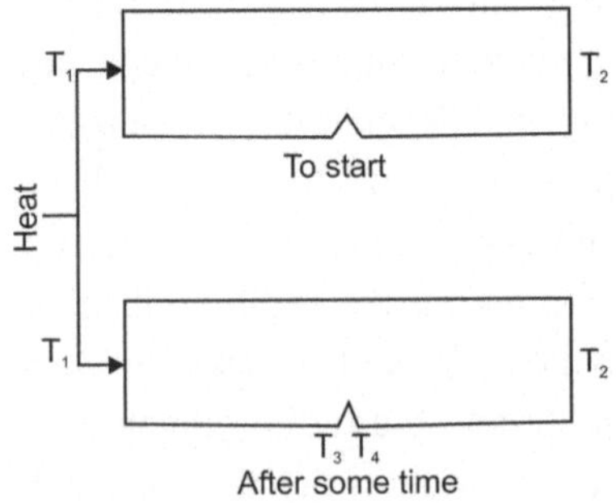

Fig. 15.2 Crack in metal offering resistance to heat flow.

Consequently, the heat distribution in the object will not be uniform. In such a case we can see from Fig. 15.2 that

$T_1 \neq T_2 \neq T_3$ at the bottom surface

$T_1 = T_2$ at the top surface

$T_1 = T_3$

$T_2 = T_4$

and $T_3 \neq T_4$ and the difference between them will be appreciable.

If we can supply heat to the object under test and study the temperature distribution, we can have a fair idea of its uniformity. This is the basic principle behind the thermal testing procedure.

Requirements

For an effective testing, we should have the following:

1. A means of effectively and quickly heating the object under test, where the heat is applied by any of the following means:
   (a) direct thermal contact with a heat source,
   (b) an electric current,
   (c) induction heating, or
   (d) infra-red heat source.
2. A good and accurate method of studying the temperature distribution.

This is the most important part of testing. The accuracy of the test is dependent upon the accuracy with which we can measure the temperature distribution. Various means of studying the temperature distribution are in vogue, such as:

(i) Wax coating.

(ii) Frosting.

(iii) Change in colour of oil.

(iv) Formation of characteristic oxides on the surface of the object.

(v) Tempilstik.

(vi) Tempilaq.

(vii) Thermocolor.

(viii) Thermocron.

(ix) Temperature sensitive phosphors.

(x) Infra-red films.

(xi) Bolometers.

(xii) Resistance thermometers.

(xiii) Thermocouples.

(xiv) Photo-conductive materials.

Some of the important and commonly used methods are explained hereunder.

Frosting the test surface

Spraying the mirror finished metal or alloy surface with a solution of 40% Diphenyl in carbon tetrachloride leaves a frosted or dull surface. When the object under test is induction heated, the region of the defect gets quickly heated up and the frosted area clears out.

Frost tests are common in testing the bond in the clad uranium nuclear fuel element. This test is however not sensitive enough to reveal defects of less than 0.25 cm width.

Phosphors

Phosphor is a highly temperature-sensitive material close to a transition temperature. The substance is mixed in a liquid plastic and applied to the surface under test. Phosphors are applied as a thin layer of 0.0025 cm thickness to the surface under test.

The efficiency of the phosphor can be best understood and its usefulness appreciated if we know that a zinc cadmium sulphide phosphor undergoes a change of 20% in its emissivity, even when there occurs a temperature difference of 2° F, on exposure to ultraviolet light.

The phosphors have different temperature ranges in which they possess the maximum sensitivity. The phosphor referred to above is sensitive in the temperature range of $5^{\circ} - 55^{\circ}$ C.

Tempilstiks and tempilaqs

Tempilstiks are patented crayons which are sensitive to temperature. Commercially, there are about a hundred types of tempilstiks in the market. These cover the temperature range of 60° C to 1300° C. Each of them is sufficiently sensitive to indicate a specified temperature with ± 1% error.

The tempilstiks are used like chalk crayons (Fig. 15.3) to mark the surface of the specimen under test. Heat is applied later and the mark liquefies.

Fig. 15.3 300° F tempilstik° in adjustable metal holder.

Tempilaqs are also patented products of M/s. Tempil, N.J., U.S. Tempilaq consists of materials of calibrated melting point suspended in an inert, volatile but non-inflammable vehicle. They too cover the same temperature range as tempilstiks and are equally sensitive (Fig. 15.4).

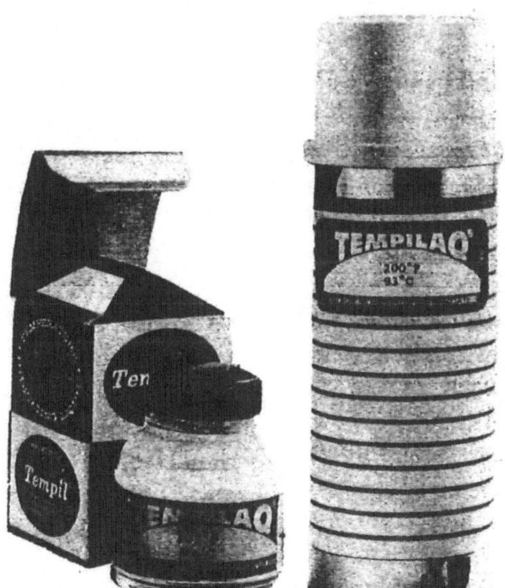

Fig. 15.4 Tempilaqs in different containers in two temperature ranges.
(*Courtesy* : Tempil, Hamilton Boulevard So Plainfield, N.J., U.S.A.)

A thin smear of the appropriate tempilaq is applied to the work piece. It dries instantly and leaves a dull, opaque mark. On heating, when the specified temperature is reached, the mark liquefies sharply.

Temperature sensitive pigments

These are a recent development in the field of thermal testing. Branded and patented pigments are manufactured in West Germany. These are known by the name 'thermocolors'. These pigments characteristically change their original colour at certain temperatures. This colour change is very marked (not just into shades of the original colour) and quite permanent for a few hours.

Some thermocolour pigments undergo more than one colour change depending upon temperature. The following is an example of one such instance:

Colour	*Temperature*
Pink	Room temperature
Light Blue	45°C
Yellow	145°C
Black	175°C
Olive Green	340°C

Advantages

These methods of testing are cheap, reasonably simple and easy to interpret. The only drawback may be their inability to detect deep seated flaws.

There appears to be good scope for research and development in these testing techniques.

CHAPTER 16

Electrical Methods

The term 'electrical methods' is very broad based. When electrical energy is passed through a metallic material, the material exhibits one or other of the following properties:

1. Electrical resistance.
2. Thermo-electric effect.
3. Static electric field.
4. Eddy currents.
5. Tribo-electric effect.

With a knowledge of the above manifestations and the way they are affected by the defects in the material, the defects, if any, can be traced.

The above techniques can be used to investigate many types of defects in metals and alloys. Almost all varieties of defects, starting from laminations in rolled sheet, insufficient and poor adhesion of bimetallic strip and cracks in electrical brushings, to poor welded and brazed joints and casting defects, can be traced by adopting a suitable technique. Besides, mixed up metal stock can be sorted out and the thickness measurements of plated and clad metals can be made by these techniques.

Electrical Resistance Method

All the metals and alloys are good conductors of electricity to varying extents and at the same time offer some resistance to the current flow. In the case of a good conductor, however, the resistance will be very low. Again, in the same metal the electrical resistance will be the same throughout only if it is completely homogeneous. If there is a defect, the resistance or resistivity will be more in that region. This is the principle underlying this method of testing.

When an electric current flows through a metal, there will be a potential difference between any two points in it. When a defect lies between them, the potential difference will be different in the region.

In a small metal specimen of uniform dimensions, the resistance between any two points, a fixed distance apart, is constant. The resistance is given by the equation:

$$R = \rho L/A$$

where R = resistance of the specimen

ρ = resistivity of the specimen

L = length of the specimen

A = cross-section area of the specimen.

The General Motors Corporation, U.S.A., employing this method of testing for finding out the cracks in camshafts, utilizes a Wheatstone bridge for determining the resistance between two points. A potentiometer or a Kelvin Bridge may also be employed (instead of a Wheatstone bridge).

The most serious practical difficulty with this method lies in tapping the surface under test for the purposes of resistance measurement. In spite of laborious and careful preparation of a clean and scale-free surface, there is bound to be a high degree of contact resistance. This may lead to recording of false values. Special electrodes and pressure contacts are developed to avoid the contact resistance. In spite of these accessories, great care is necessary in this method of testing, for the directions of the tappings should be exactly at right agles to the current flow. Though the established principle is that the current passes through the shortest possible distance of the matter, the bulk of the part under test is likely to give erroneous values of potential drop. Thus, the thinner the part under test, the more accurate the testing would be.

THERMO-ELECTRIC METHODS

All metallurgy students, rather science students, are familiar with the Seebeck and Peltier effects. Two dissimilar metal wires or strips are welded at one end to function as the hot junction. The other ends, both being maintained at the room temperature, form the cold junctions. Across them is connected a milli-voltmeter to measure the e.m.f. generated, when the hot junction is heated. The electro-motive force which produces this current is called the thermal electro-motive force. This thermal e.m.f. will vary with the temperature difference between the hot and cold junctions and the materials. Such a combination of dissimilar metals is called the thermocouple.

The instrument shown in Fig. 16.1 consists of two probes, both made out of the metal or alloy (thermocouple materials). It is used in finding out the temperature of a body or the temperatures at different places in the same body. In the thermal method of testing, we have seen that the defective region in the object under test has a different temperature than the rest of the object. By maintaining one of the probes of the instrument, i.e., cold junction, at a constant temperature and contacting the object under test with the other probe, i.e., hot junction, the temperatures of the various regions can be found out. The milli-voltmeter connected in the circuit being always calibrated in °C, any fluctuation in the temperature when the probe touches a particular spot on the object under test, indicates that it is a defective region.

Fig. 16.1 Thermocouple.

EDDY CURRENT METHODS

When a current-carrying conductor coil is brought near a metal specimen, eddy currents are introduced in the metal. The reason for this is electromagnetic induction. The current in the coil magnetizes the metal and when it is moved, the magnetic lines of force are cut and current flows in the metal.

The magnitude of the induced eddy current is related to :

(i) the magnitude and frequency of the alternating current flowing in the coil,

(ii) the electrical conductivity of the specimen,

(iii) the magnetic permeability of the specimen,

(iv) the shape of the specimen,

(v) the relative positions of the coil and the specimen, and

(vi) the metallurgical nature of the specimen.

When a coil carrying alternating current is brought near a metallic object, eddy currents are induced in the object by electro-magnetic induction. If the current flowing in the coil is I and is of a constant amplitude, it sets up an alternating magnetic field H_c about it. Since the metallic object in the vicinity of the coil is also a conducting material, this alternating magnetic flow will induce an eddy current flow I_{cc} in the object (see Fig. 16.2). The induced eddy current flow will in turn generate a magnetic field H_{cc} in a direction opposite to the primary field H_c. Thus, the primary field is weakened and the inductance of the coil reduced. The presence of a defect in the metallic object will modify the local eddy current flow and these small changes in the eddy current flow are detected by using suitable instruments.

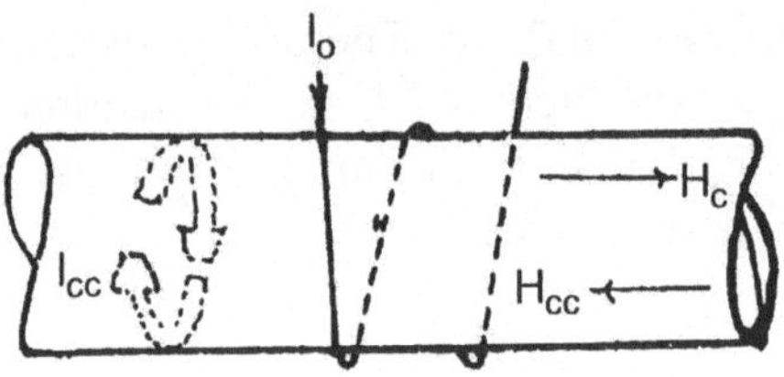

Fig. 16.2 A test coil assembly.

The induced eddy currents are generally concentrated near the surface of the object resulting in the so-called "skin effect". In the case of a plain conductor the current falls off exponentially with the depth below the surface. The relation is mathematically expressed as:

$$d = \frac{1}{[\pi f \mu \sigma]^{1/2}}$$

where d = the depth of penetration in metres,

f = frequency in cycles per second,

μ = magnetic permeability of the material in Henrys per metre, and

σ = volume electrical conductivity of the materials in mhos / metre.

The depth of penetration in a uniform field is the depth at which the current is equal to $1/e$ times the value at the surface. This roughly works out to be about 37 % of the surface value.

In order to locate sub-surface defects or those at a sufficiently deeper position in the object, it is essential that the depth of penetration of the current is more. Bareham has derived expressions for the distribution of eddy current density and phase within a tube material. It is shown that the depth of the eddy current penetration into the tube wall decreases with an increase in the value of

$$\sqrt{\omega.\,\mu.\,\sigma.\,b}$$

where ω = $2\,\pi f$,

f = test frequency in cycles per second,

σ = electrical conductivity,

μ = magnetic permeability, and

b = outer radius of the tube.

Thus for a tube or solid of a given material and dimensions, a higher test frequency is chosen for surface or near surface inspection, and lower frequency is employed for probing through the section.

Detection of the Eddy Currents

As mentioned above, the only method suitable for noting the eddy currents is by the measurement of impedence. This presents certain difficulties since the impedence value is affected when the specimen-coil spacing changes. These effects are likely to mislead the observer into thinking that they are defects. Fortunately, in a majority of the instances, such effects can be discriminated by utilizing the phase or time relation or the amplitude of the eddy currents.

The test coils are classified into two types:

1. Absolute coils which directly measure inductance.
2. Differential coils where pairs of coils are used, connected in series and opposing each other.

A standard specimen is surrounded by one coil and the object under test is surrounded by another. When the output from the two coils is zero, the object is a sound one. Otherwise, there is a defect.

Three types of probes are in common use. The connection diagrams (Fig. 16.3) give a comprehensive idea regarding the electrical nature of the testing. The concentric coil completely surrounds the specimen. It probes an annulus of the specimen whose width is equal to the effective width of the coil. The second type, known as the point probe, consists of a small coil that can be placed near the surface of the specimen. It probes an area essentially equal to the cross-sectional area of the probe. The third type of coil, the inside or bobbin coil, is used for testing tubes or pipes.

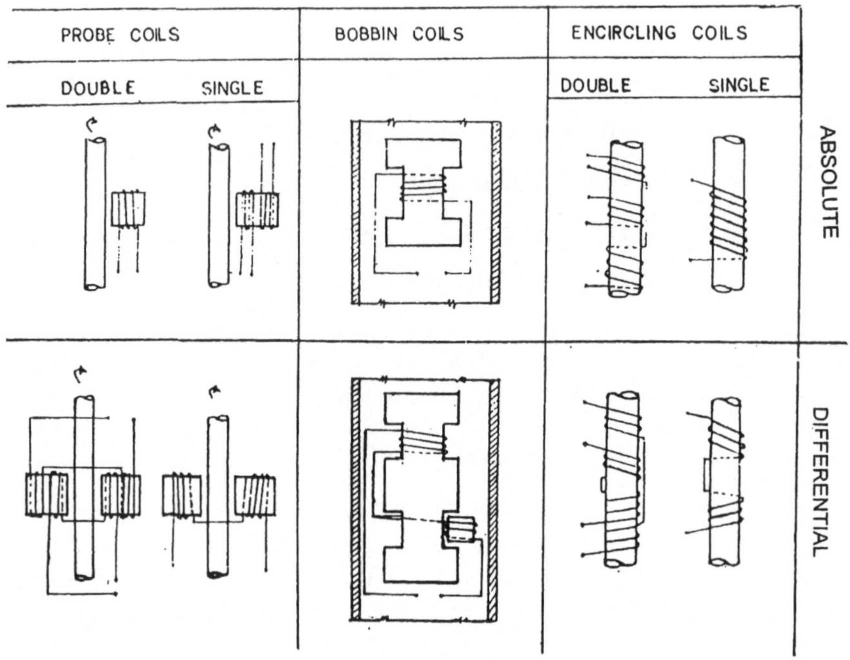

Fig. 16.3 Types of coils.

The effect of the defects on the impedence of a coil is not so simple as to facilitate an easy and accurate measurement. The common sources of error would be the inductance of the lead itself (in the case of tiny coils) which will be more than that of the test coil. Further, at high frequencies, the distributed capacitance and the stay-lead capacitance will greatly effect the test results. The size and type of the test coil are thus governed by:

1. the material being tested,
2. the nature of the defect being looked for, and
3. the size of the defect.

It the overall conductivity of the material is to be measured, a cylindrical coil is employed, provided the material is cylindrical. Again, a sufficiently low frequency is selected to ensure a reasonable penetration of the currents.

If the defect being probed is a surface or sub-surface one, a small flat coil assembly is very convenient. It should, however, be noted that for a coil of a given diameter, the sensitivity to a particular defect decreases rapidly as the size of the defect becomes less than the diameter of the coil. There is no comprehensive theory for the operation of probe type coils.

Eddy Current Instruments

The eddy current instruments are used to detect the surface defects in tubes, rods, wires, etc., mostly of non-magnetic materials. The most commonly tested metals include copper and stainless steel. If ferrous materials are to be tested, they should be magnetically saturated by using special auxiliary equipment (Figs. 16.4 and 16.5).

Fig. 16.4 Magnetic saturation unit for ferromagnetic bar and tube up to 37 mm.
(*Courtesy* : C.N.S. Instruments Ltd., London, U.K.)

Fig. 16.5 Magnetic saturation unit for inspection of ferromagnetic tubing up to 100 mm diameter.
(*Courtesy* : C.N.S. Instruments Ltd., London, U.K.)

Various eddy current instruments are in the market with different brand names. Suitable equipment can judiciously be selected depending upon the type of the work to be inspected and the technical and financial advantages offered.

The CeNtest type 607 is a probe type instrument and is semi-automatic (Fig. 6.6). The eddy current probe is mounted on an electrically driven displacement actuator. The metallic object under test for surface defects is placed within the range of the actuator and the probe senses the surface of the object and feeds back. The instrument being designed to detect surface defects, operates on high frequencies (nearly 10 MHZ). Further, the changes in the probe reactance are independent of the resistivity of the surface layers due to the high frequencies employed. The reactance will be solely dependent upon the gap between the surface under test and the tip of the probe. The operation of the instrument falls into the category of the aforesaid 'absolute coils.' Other instruments of similar type are the Probolog, Magnatest, Defectometer, Sigmatest, etc.

Fig. 16.6 Eddy current probe (P) inspecting sheet.
(*Courtesy* : C.N.S. Instruments Ltd., London, U.K.)

Continuous Inspection and Testing

Equipment capable of detecting longitudinal splits, cracks, seams, inclusions and, to some degree, any transverse cracks on the bar and tube stock, is available. One such popular equipment is the CeNtest types 710 (and also 730), the former incorporating a quick response recorder and the latter incorporating a surface inspection unit as well, Fig. 16.7. These machines can be fitted in the production line itself and are designed to function up to rolling speeds of about 200 meters per minute. The oscilloscope unit built in the instrument detects the defects and actuates the recording mechanism.

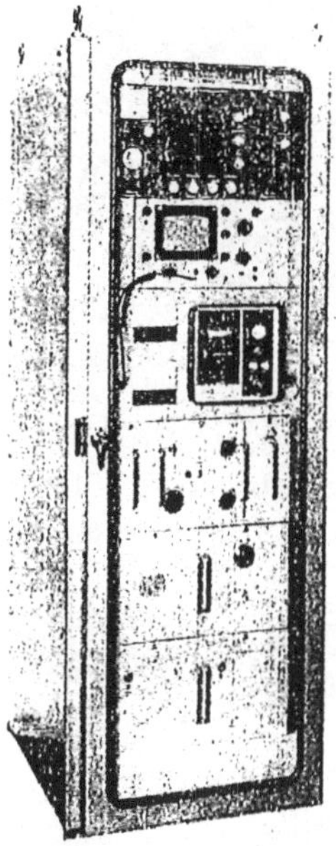

Fig. 16.7 Magnetic sorting bridge (ceN test 710).
(*Courtesy* : C.N.S. Instruments Ltd., London, U.K.)

Figs. 16.8, 16.9 and 16.10 show the automatic inspection units fitted in M/s. Osborn Steel Limited in their mill at Ecclesfield. Sheffield. The whole installation comprises the feeding tables in the vibrators seen to the left in Figs. 16.8 and 16.9 and to the right in Fig. 16.10. The central inspection section comprises of two magnetic saturation units, a smaller one for bar up to $\frac{1}{2}$" (Figs. 16.8 and 16.9) and the larger saturation unit (Fig. 16.10). The unit seen at the left of Fig. 16.10 is a demagnetization unit to remove all the residual magnetism in the final product.

Fig. 16.8 Automatic inspection unit fitted in the production line.
(*Courtesy* : C.N.S. Instruments Ltd., London, U.K.)

Fig. 16.9 Automatic inspection unit fitted in the production line.
(*Courtesy* : C.N.S. Instruments Ltd., London, U.K.)

Fig. 16.10 Automatic inspection unit fitted in the production line.
(*Courtesy* : C.N.S. Instruments Ltd., London, U.K.)

Instruments of the above type similar to Fig. 16.7 but with higher frequency operation can also be used to test the weld quality of the seam welded stainless steel tube. Unlike copper, the scrap value of stainless steel being much lower when compared to the prime material, it will be an asset to be able to locate the faulty weld immediately after it occurs, so that the operator can suitably correct the defective mechanism and avoid losses.

Detection of Ferrous Inclusions in Non-Ferrous Wire

Fine copper wire is cold drawn on multi-die machines. One great difficulty experienced in the process is the breakage of the wire. Apart from the die design and operational defects like improper lubrication, which can be controlled, an important drawback is thought to be the ferrous inclusions. The British Non-ferrous Metals Research Association has designed a ferrous inclusion detector. This eddy current instrument inspects, on the production line itself, wire sizes up to 2.5 mm diam at drawing speeds between 5 and 25 metres per second. Ferrous inclusions of the minimum size of 0.15 mm or clusters of that size are detected by this instrument, and thus early warning is given regarding the contaminated wires.

The mechanical arrangement of the detector (shown in Fig. 16.11) essentially consists of a permanent magnet and a coil through the core of which the wire passes. As the magnetic field is parallel to the wire axis, no primary effects are produced due to the movement of the wire. Lateral vibrations of the wire are minimized by suitable idlers through which the wire is guided. Any ferrous inclusion in the wire passing through the coil disturbs the magnetic field and is instantly detected by the coil. The output signal drives a chart recorder or counter thus indicating the number of inclusions in selected size ranges. The only disadvantage of this unit is its inability to detect non-ferrous inclusions.

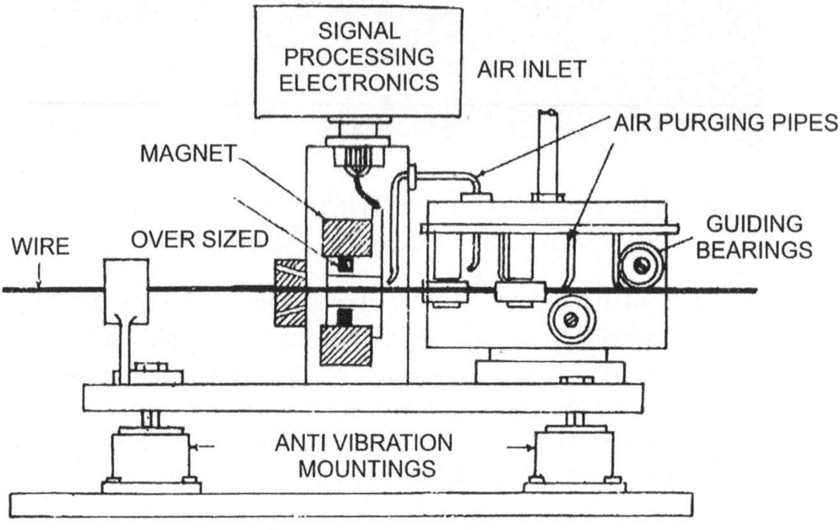

Fig. 16.11 Sketch of the B.N.F. ferrous inclusion detector.
(*Courtesy* : C.N.S. Instruments Ltd., London, U.K.)

Thickness Measurements

The photograph of an eddy current thickness tester (Eddy tester Model 173) is shown in Fig. 16.12. The instrument consists of a coil which is energized with an alternating current. The magnitude of the eddy currents thus produced is largely dependent on the electrical resistivity of the metal. This instrument assesses the change in the characteristics of the coil electronically and gives a measure of the proximity of the metal to the coil. Thus, the thickness of the coating lying intermediate is measured directly on the scale of the instrument. (Fig. 16.13 shows some autoparts inspected by the Eddy current techniques.)

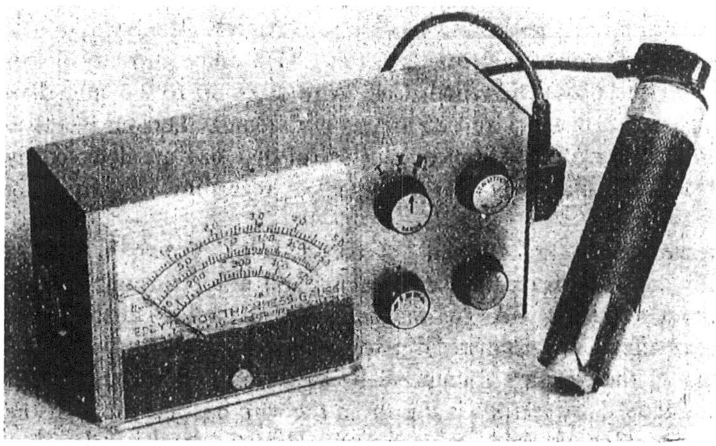

Fig. 16.12 Eddy tester thickness gauge.
(*Courtesy* : Elcometer Instruments Ltd., Manchester, U.K.)

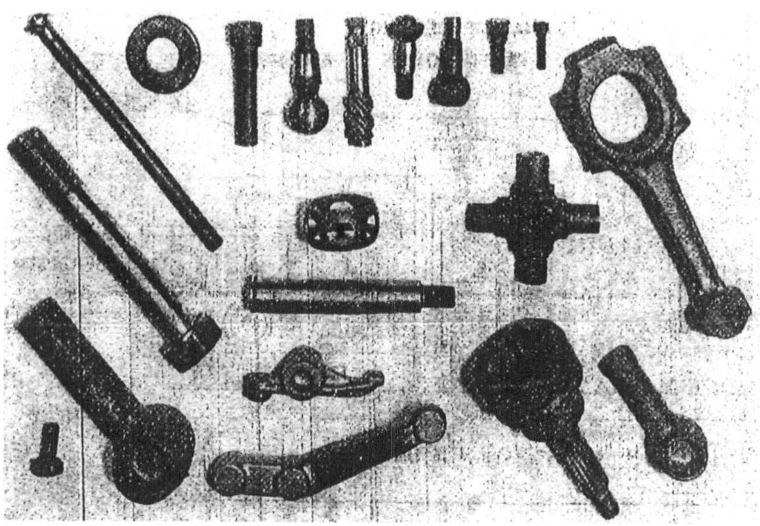

Fig. 16.13 Some autoparts inspected by Eddy current techniques.
(*Courtesy* : C.N.S. Instruments Ltd., London, U.K.)

Various types of machines are available for measuring the thickness of anodisings, protective and decorative paints, enamels and plastics, besides platings.

TRIBO-ELECTRIC EFFECT

A voltage is generated when two metallurgically dissimilar metals are 'agitated'. The magnitude of the voltage thus generated ranges from a fraction of a microvolt to several millivolts, depending upon the two metals/alloys which are in contact. This phenomenon is known as "triboelectric effect".

The redistribution of the electrons in the two dissimilar metals consequent to the agitation is the reason for this tribo-electric effect. These redistributed electrons produce a sizeable electric current. All the metals can be arranged in "tribo-electric series". In such a series, a higher metal is positive with respect to a lower one. For example, suppose the series consists of A, B, C, D, E, F, etc., then the metal A is positive with respect to B, the metal B is positive with respect to C and so on. Again, the metal A will be more positive when compared to C than when compared to B.

Tribo-Electric Series

Mg	Fe
Pb	Ni
Al	Sn
Mn	Ag
Zn	C
Cd	Brass
Cr	Pt
	Au

Much work needs to be done in the complete utilization of this effect for testing purposes. However, this does not appear to be difficult due to the following reasons:

(i) The friction necessary to produce this effect need not be so high as to cause any surface rupture.

(ii) The contact of other dissimilar metals with those under test does not effect the voltage generated in any way. Thus, actual measurement is not a problem.

(iii) Surface roughness does not influence the tribo-electric effect.

(iv) Variations in section or size have no influence on the tribo-electric effect.

The only condition to be strictly adhered to is that the surfaces of the metals under test should be thoroughly clean, and be completely free from rust, scale, grease, etc.

CHAPTER 17

Surface and Thickness Measurements

Besides just being the dimensional check-up which it was sometime back, this branch of non-destructive testing has acquired variety and become the most common nowadays. In these days of quality consciousness a hundred percent inspection becomes necessary in many cases. Thus, the duties of the control laboratory and the inspection department in an engineering concern are overlapping increasingly.

With the progress made in speed travel and particularly space research work, high speed technology is the goal of every scientist and technician. High speed engines require the use of bearings and shafts of very smooth surfaces. Degrees of finishings in terms thousandths of an inch are gone by and time has come when we should talk in terms of microns.

Electroplatings, claddings and vapour coatings are the jobs of the metallurgist. Having made a particular sophisticated finish, it will be embarrassing to rely upon someone else to check up the degree of perfection. Besides knowing how to measure the thickness, it is equally desirable that the metallurgist has an idea regarding the surface measurements. Details of both the techniques are generally omitted. But at the same time, making an effort to describe the basic principles involved in these techniques is not out of place. That is precisely what is done here, in this chapter.

SURFACE ROUGHNESS AND ITS MEASUREMENT

At the outset we may say that no surface is smooth. A machined surface is smoother than a cast surface. A ground surface will in turn be smoother than a machined surface. Again, a scraped surface will be smoother than the ground surface. Lastly, a buffed surface will still be smoother. Thus, there is no strict limit to the word 'smooth' except in a sense of comparison.

A smooth or plain surface means that it is not made up of visible or feelable ridges. Microscopic examination of a surface will reveal the degree or amount of its unevenness. The more is the magnification required to find a surface irregularity, the smoother it is. In other words a smooth surface is one on which the points of highest level and those of the lowest level do not differ by any great amount.

Instruments

The straight edge

Straight edge is a simple instrument. The degree of accuracy in the measurement is in terms of millimetres only. The instrument consists of a steel bar with one or more straight edges. The surface being examined is painted and the straight edge is placed against the specimen. The regions where paint is removed can be taken as those which are projecting.

The straight edge is used along with a feeler gauge to find out the depths of troughs on the surface.

Surface gauge

This is actually a scribing instrument. The instrument consists of a firm base on which rests a vertical post. The vertical post can be fixed in any desired position and clamped by using the knurled screw (1). The scriber too can be fixed at any height and position on the post by working another knurled screw (2) as shown in Fig. 17.1.

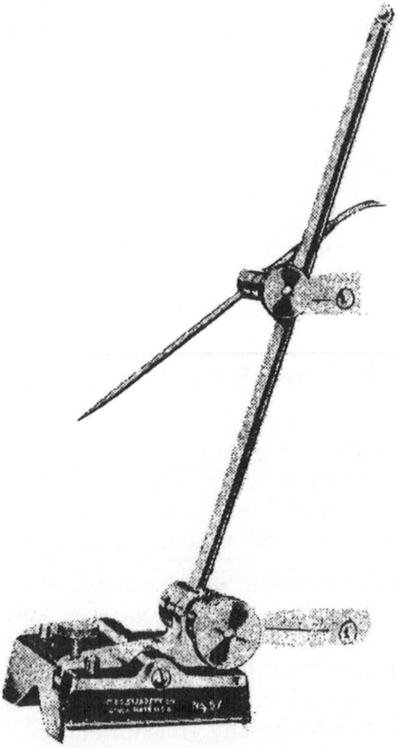

Fig. 17.1 Surface gauge.

(*Courtesy* : L.S. Starett Co., Athol, Mass, U.S.A.)

The surface to be checked is coated with chalk and placed on a surface plate and the instrument is taken near it. The scriber is positioned to the required height. The instrument is moved on the plate so that the scriber moves over the surface under test. Projections are felt by the scriber and the chalk is removed therefrom.

Optical flat

This is based on the principle of interference of light. Optical flats are 'too highly polished natural quartz plain lenses'. Each of them has a thickness of about 1.6 cm and a diameter of 5 cm. The two surfaces of the flat are absolutely parallel.

The optical flat is placed on the plain surface under test and light (monochromatic) is focused on it. The light is reflected by both the optical flat and the surface under test through the thin layer of air between them. Even if an air film of the thickness of half the wavelength of light is present between them, interference occurs. This produces interference bands which appear as alternate white and dark lines (Newton's rings type effect).

When the surface under test is perfectly plain, these bands are parallel to each other. If not, these appear as contours.

Though the optical flats of 5 cm diameter are commonly used, optical flats of up to 25 cm diameter are available. Generally fluorescent helium light is employed. The instrument can also be used for comparison measurement, warranting precision up to a few Angstroms.

Profilometer

Profilometer is a direct reading instrument, and functions based on a simple principle. The surface is probed by a diamond point tracer. The machine consists of a tracer which converts the up and down movements of the tracing point into electrical voltages. A motor drive operates the tracer and is called the mototrace. The voltage fluctuations are amplified and fed to a screen. The screen can be read in hundredths of millimetres or the values recorded on a sheet.

The contour obtained is divided into vertical strips over a reference line AB (Fig. 17.2). The heights or depths are measured for each strip 1, 2, 3, etc. A mean value or a root mean square value is calculated from this and it is taken as the average surface roughness.

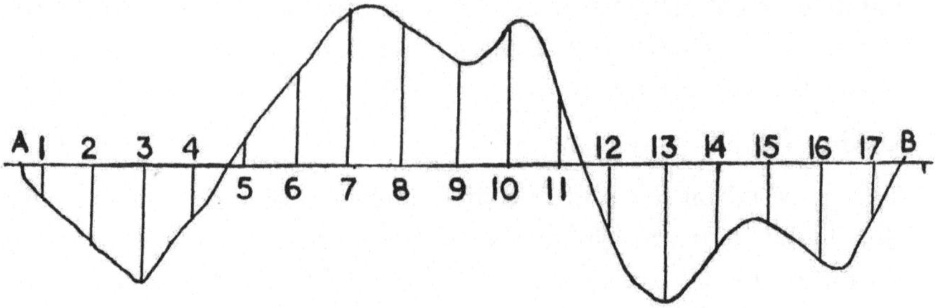

Fig. 17.2 Contour obtained from a profilometer reading.

Strip. No.	Depth/ht	$(\text{Depth/ht})^2$
1	3	9
2	12	144
3	19	361
4	6	36
5	2	4
6	17	289
7	22	484
8	20	400
9	17	289
10	22	484
11	10	100
12	10	100
13	18	324
14	11	121
15	9	81
16	12	144
17	9	81
	219	3451

$$\text{Arithmetical average} = \frac{219}{17} = 12.88$$

$$\text{Root mean square average} = \sqrt{\frac{3451}{17}} = 14.24.$$

THICKNESS MEASUREMENT

A great variety of sheet metal work which is produced needs accurate control of the thickness. Besides, precision coatings and plating should conform to rigorous thickness specifications especially when needed for sophisticated scientific purposes. The usefulness is so vital that the material apart from being useless, if scrapped, becomes a big financial loss.

Thickness measurement is generally performed with two types of gauges–contact and non-contact.

The contact type gauges include the ultrasonic resonance, electrical conduction, magnetic and electromagnetic gauges. These are suitable for all the measurements connected with the sheet, but become useless in particular instances like measurement at higher temperatures, etc. In such cases, the non-contact type instruments are used. Radiation gauges are the recent development in the sphere and they are capable of recording thickness with an accuracy of 0.1 to 0.2%.

Contact Gauging

Ultrasonic gauges

It is seen under ultrasonic testing that, when an ultrasonic pulse is fed into an oscilloscope after passing through a metallic object, two peaks are observed on the two extreme ends of the oscilloscope screen. They correspond to the reflections of the waves from both the surfaces of the specimen. The distance between the two peaks on the screen can be computed and interpolated to give the thickness of the metallic object. This gives us a fundamental concept of how the thickness measurement can be made by employing the ultrasonic waves.

In actual practice longitudinal waves are utilized. These waves are passed through the sheet whose thickness is under investigation. These waves cause the sample to vibrate in the direction of its thickness. The frequency employed is the resonance frequency, i.e., the oscillation is at the maximum intensity. Thus, the amplitude of vibrations in the crystal is increased resulting in a greater induced voltage. At one of the natural frequencies of vibration of the specimen in its thickness direction, for which the thickness of the sample is equal to an exact number of half wavelengths, resonance occurs. If the frequency is f_n and the corresponding wavelength is $\lambda\, n$, the thickness t is given by $t = n\, \lambda\, n/2$, where n is a whole number.

The frequency is given by

$$f_n = c/\lambda\, n = nc/2t,$$

where c represents the velocity of the sound in the specimen and f_n is the nth harmonic of the frequency f of the specimen. Thus, $f_n = nf$.

Again, by increasing the frequency, suppose the resonance occurs again at $(n + 1)$th harmonic, the frequency is $f(n + 1)$

We have, $$f(n + 1) = (n + 1) f = (n + 1).c/2t.$$

Subtracting,

$$f(n + 1) - fn = f = c/2t.$$

or $$t = \frac{c}{2f}.$$

In the actual determination process of thickness, a probe with a fundamental frequency of 1.5, or 2 or 2.5 Mc/s. is used. The resonance frequencies are observed and the frequency of vibration of the solid is obtained by subtracting the frequencies observed at resonances from the next higher resonance frequencies. Knowing the velocity of sound in the particular metal or alloy C from standard charts, the thickness of the sheet t can be computed. Values of C for some common metals and alloys are given in the table below.

Sl. No.	Metal/alloy	Velocity of sound Mts/S
1.	Aluminium	6400
2.	Brass	3500
3.	Copper	4700
4.	Iron	5900
5.	Lead	1200
6.	Nickel	5600
7.	Steel	6000

Electric resistance gauges

In a sheet metal, the electrical resistance between two points on the surface is inversely proportional to the thickness. That is, for a particular value of the voltage the current flow is proportional to the thickness.

An electric current is introduced into the sheet through a pair of electrodes. The potential drop is measured by another set of electrodes called the potential electrodes.

Limitations

(i) The process is not suitable when the sheet is thick. The current-thickness relationship does not hold good for thick stock.

(ii) The measurements cannot be made at edges.

Magnetic induction gauges

A magnetic yoke placed on the specimen is excited by a low current of approximately 6 amps fed from a storage cell. The steel specimen gets saturated with magnetism. When the current is put off, the magnetic field breaks down. This gives a pulse and it can be read by means of an accurate galvanometer (ballistic type). The thickness of the material can be found out from the deflection of the galvanometer since the magnetic flux induced and the consequent pulse when the magnetism has left is proportional to the material thickness.

Magnetic attraction gauges

The force necessary to pull out a small magnet from a coated surface is proportional to the thickness of coating. If the coating is magnetic, the force necessary to pull out the magnet will be more if its thickness is more. On the other hand, if the base is magnetic and the coating non-magnetic, the force required to pull out the magnet decreases with increasing thickness of the coating.

The photograph of an Inspector Thickness Gauge (Model III) is shown in Fig. 17.3. The instrument uses the attractive force of a permanent magnet to a ferrous base as its method of measurement.

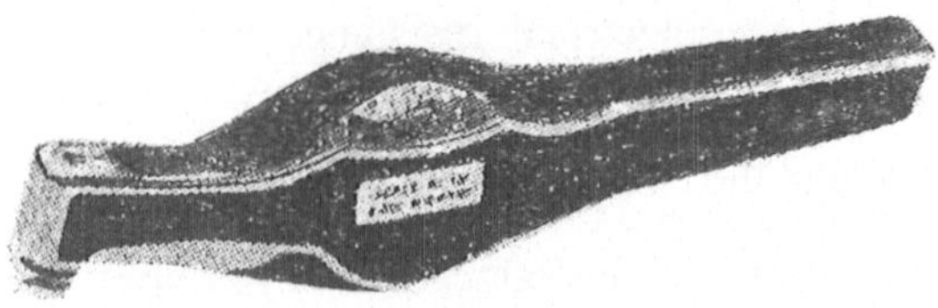

Fig. 17.3 Magnetic attraction gauge.
(*Courtesy* : Elcometer Instrument Ltd., Manchester, U.K.,)

In the process of use, the scale wheel is pushed forward, allowing an internally mounted friction arm to place the magnet onto the surface of the coating. By pulling back on the wheel, a coiled spring is compressed which applies a pull-off force to the magnet. At the point of break the thickness reading is taken. An adjustable zero is incorporated for greater accuracy.

Magnetic Reluctance Gauging

In a transformer, the voltage in the secondary is proportional to the flux which links it with the primary winding. Again, the flux is proportional to the magnetic reluctance of circuit.

The plating thickness of nickel (a ferromagnetic material) on nonmagnetic bases is commonly measured by this method. The object forms part of the magnetic circuit. Under such circumstances the secondary voltage is dependent upon the magnetic reluctance of the nickel plating. From the elementary principles of magnetism, we know that the magnetic reluctance of a body is directly proportional to the 'path length' and inversely proportional to its area of cross-section. In the case of a plated object the width and length are constant, out of the three dimensions in consideration. Hence the magnetic reluctance is inversely proportional to the nickel thickness, i.e., the plating thickness.

Non-contact Gauging

Measurements by radiation absorption

X-rays, γ-rays and alpha and beta particles are energy radiation. They are absorbed by the matter through which they pass.

The material whose thickness is to be determined is placed between a source of X- or γ-rays and a detector. The amount of absorption, which obeys a particular relationship with its thickness, is detected.

Alpha and beta particles are emitted when certain nuclei undergo radioactive decay. An alpha particle is a helium atom whose orbital electrons are removed. These particles are not powerful in penetrating matter as X-rays or γ-rays. When they travel through matter their energy is lost due to the ionization process. They have a certain range. This range is expressed in terms of centimetres of air at 76 cm of mercury and 15.5°C.

When discussing the penetration of matter, the term 'stopping power' is used. The stopping power is defined as the reciprocal of that thickness of the material whose ability to stop the alpha particles is equal to 1 cm thickness of air. The stopping power of the materials has no definite relationship with respect to the velocity of the alpha particles. However, it shows that the alpha particles of medium range obey Geizer's formula as follows:

$$R = a\, v^3$$

where R = range in cm,

a = 9.6×10^{-28} sec^3/cm^2, and

v = velocity of the alpha particles in cm/sec.

By employing alpha particles instead of X-rays or γ-rays and using an ionization chamber to measure the fall in their energy, the thickness of materials can be measured.

Beta particles are high speed electrons. They possess more energy of penetration than the alpha particles. There is an empirical relationship to compute the range of the beta particles.

$$\rho t = 0.54\, Em - 0.16$$

where t = thickness of the absorbing material

ρ = density of the absorbing material (gm/cm^3)

Em = maximum energy of the beta particles in MeV.

By virtue of being very tiny, these particles do not travel in straight lines through the absorbing medium. It can be assumed that they follow a random or zig-zag path.

Beta rays are detected by the use of an ionization chamber or a betarays detector.

These methods employing alpha and beta particles are particularly suited for the measurement of the wall thickness of tubing and their roundness.

Fluorescent method

When a metal is irradiated with photons of sufficient energy, part of it is absorbed by the metal and a part re-radiated. This phenomenon of re-radiation is referred to as fluorescence.

Fluorescent radiation has a characteristic wavelength depending upon the material which fluoresces. This is measured by fluorescent X-ray spectro-graphical techniques to reveal the thickness of the material.

Eddy current techniques

Various methods based on the differences in the electrical conductivities of the coating and the base material are developed. These are outlined under eddy current techniques elsewhere in this book.

References

Manual of Radiography, Eastman Kodak Co., Rochester, N.Y., 1939.

Tasker, H.S., *Handbook of Industrial Radiology*, Ed. J.A. Crowther, Edward Arnold, London, 1949.

Symposium on Ultrasonic Testing; *A.S.T.M. Spe. TechPublications*, 101, 1951.

Mc. Master, R.C., *Non-destructive Testing Handbook*, Vols. I & II. The Ronald Press, New York, 1959.

Hinsley, J.F, *Non-destructive Testing*, Mecdonald and Evans Ltd., London, 1959.

McGonnagle, W.J., *Non-destructive Testing*, McGraw-Hill Book Co., New York, 1961.

CHAPTER 18

Defects

A motor cyclist who has learnt driving with no difficulty and has driven his vehicle for quite some years without any accident has no reason to think about an accident or its harmful consequences. But, once he comes to experience even a minor one, his self-confidence will be shaken to its roots. Thus, as long as we are ignorant of a harmful effect, we will not bother much about it. But, once we realize that there is a possibility for such an occasion to arise, we start getting a bit nervous. When such harmful effects or defects are numerous, we may even wonder whether it is really possible to avoid them totally.

Defects in metals are numerous. We should, in fact, be proud that we are able to manufacture many engineering and other items successfully. Of course, some of the defects, though present, may not be really harmful from the point of view of the performance of a particular product. Yet, defects are defects and efforts should be made to avoid them altogether.

The various metallurgical defects can be classified into three major types.

1. **Inherent defects**

   These are introduced during the initial stages of production of the metal. They are carried on through the various fabrication processes and manifest in the same or some other form in the finished material, e.g., a blow hole in the casting of an ingot. If this does not get 'welded up' during the rolling process, it manifests itself as a seam in the final rolled section.

2. Processing defects

These defects are introduced during the processing or fabrication of a product. A majority of the defects fall into this category, e.g., blow holes in castings, laminations in rolled stock.

3. Service defects

These defects are caused in the material by virtue of its service, e.g., fatigue cracks, and pitting.

The processing defects can be further sub-divided into:

1. the casting defects, and
2. the fabrication defects.

The casting defects are listed as follows:

1. Pipe
2. Blow
3. Burning-on
4. Metal penetration
5. Dirt
6. Core rise
7. Core shift
8. Fin
9. Core wash
10. Hot tear
11. Misrun
12. Vein
13. Scab
14. Buckling
15. Fissure
16. Rough surface
17. Entrapped gas
18. Sand inclusions
19. Oxidized metal
20. Unfilled mould
21. Cold shut
22. Cold shots
23. Unsoundness or porosity
24. Drop
25. Mould cut or wash
26. Rat tail
27. Shrink
28. Inclusions
29. Segregation.

Photographs of some of the above are shown in Figs. 18.1 to 18.5.

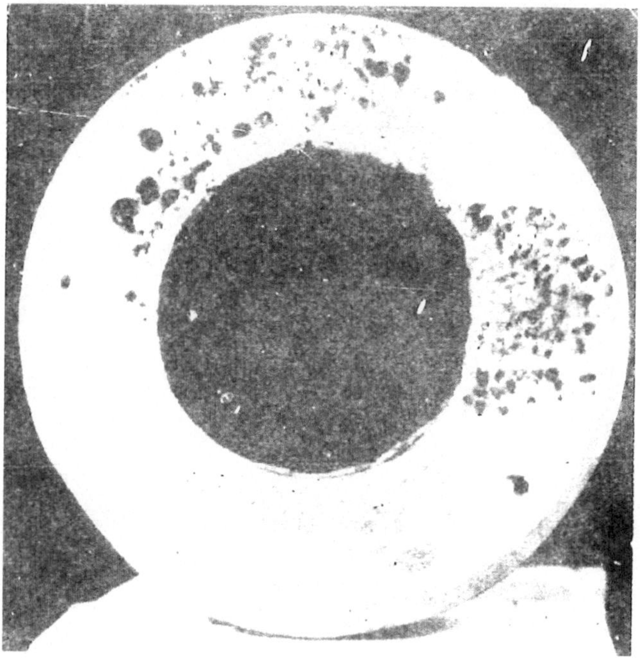

Fig. 18.1 Gas unsoundness in a casting.
(*Courtesy* : Indian Copper Information Centre, Calcutta.)

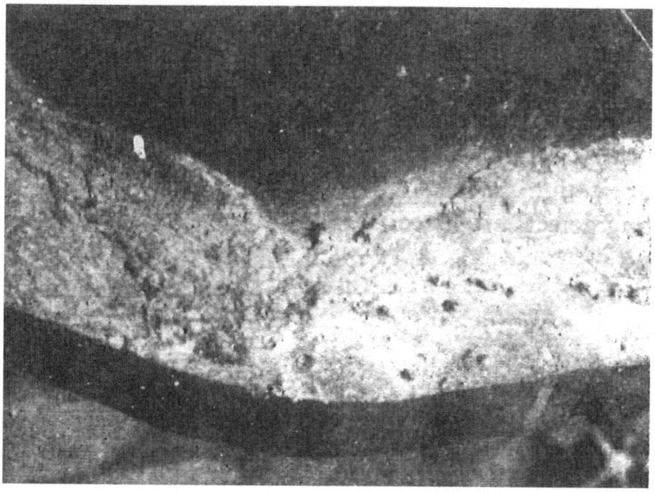

Fig. 18.2 An illustration of scab.
(*Courtesy* : Indian Copper Information Centre, Calcutta.)

Fig. 18.3 Shrinkage cavity.
(*Courtesy* : Indian Copper Information Centre, Calcutta.)

Fig. 18.4 A cold shut defect.
(*Courtesy* : Indian Copper Information Centre, Calcutta.)

Fig. 18.5 Misrun.

(*Coutesy* : Indian Copper Information Centre, Calcutta.)

Fabrication defects are listed below :

1. Lamination
2. Porosity
3. Improper joint
4. Scale
5. Fin and over-fill
6. Under-fill
7. Sliver
8. Lap
9. Seam
10. Fine cracks and roll marks
11. Scratches
12. Rolled-in oxide
13. Buckle and kink
14. Burned steel
15. Hook
16. Camber
17. Pipe
18. Shear distortion
19. Extrusion defect
20. Surface cracks
21. Cold shut.

SEGREGATION

Segregation involves the variation in composition from one region to another, in the same material. This defect is confined to alloys only. Segregation is of three types:

1. Normal segregation,
2. Gravity segregation, and
3. Inverse segregation.

Normal Segregation

Normal segregation is the variation in composition in a casting from the surface to the centre, the high melting constituent being concentrated at the surface which is the first solidifying region. In the equilibrium diagram illustrated (Fig. 18.6) it can be seen that the binary alloy containing X % B, has a solidifying range T_1–T_2. Solidification starts at T_1, when the solid separating will have only x % B, i.e., it will be A-rich.

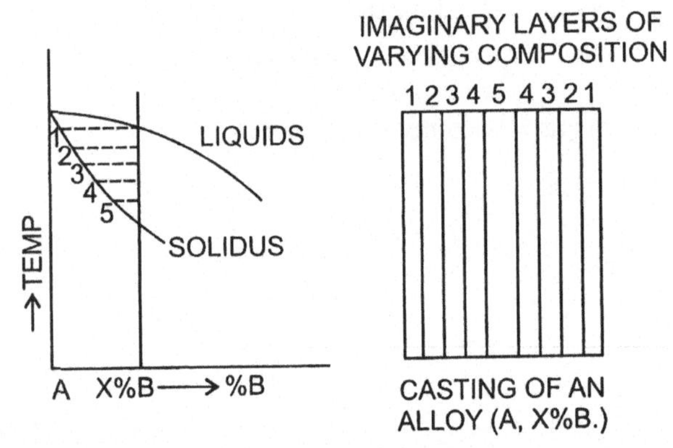

Fig. 18.6 Normal segregation (illustrated).

As the solidification proceeds with the fall in temperature, the solid separating changes its composition along the solidus line in the T_1–T_2 range. The solid gets richer in B. We know that when the liquid metal is poured into a mould, the coldest region will be that near the mould walls. The solidification commences from the mould walls and proceeds into the interior. Finally, the casting obtained will exhibit a gradual composition gradient (starting with x % B to X % B) from the surface to the centre. This phenomenon is called the normal segregation.

Gravity Segregation

Alloys in which one constituent is comparatively heavier than the other/others exhibit this segregation, e.g., lead-tin, lead-antimony, bearing materials containing lead, etc.

The mechanism behind the phenomenon is the settling of the lead rich constituents which solidify first, under gravity. Consequently, the bottom regions show an increased percentage of the heavier constituent (Fig. 18.7).

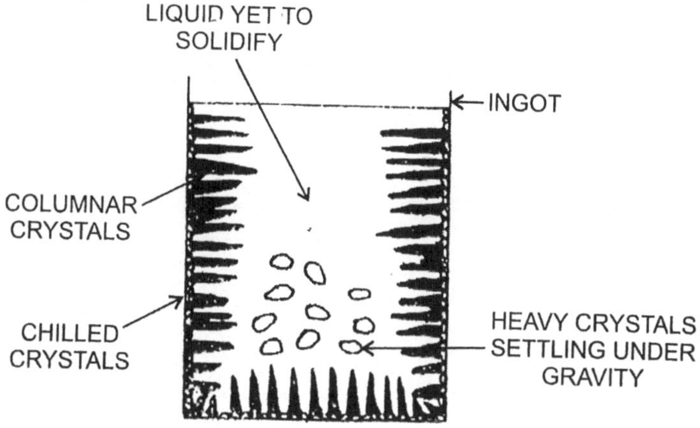

Fig. 18.7 Casting of a lead antimony alloy (a few minutes after pouring, still partly liquid).

Inverse Segregation

As the name itself suggests, this is just opposite to the normal segregation effect. In a casting, the outer regions can be found to be rich in the low melting constituent instead of the high melting constituent. There are many theories put forward to explain this phenomenon, of which the contraction pressure theory sounds most reasonable. According to this, the first formed solid, on contraction, exerts a high pressure or thrust on the yet-to-solidify internal liquid. The liquid which is rich in the low melting constituent and which is at the centre until then, being unable to flow anywhere as it is surrounded by the solid metal shell, 'reacts'. Consequently the film of solid around the liquid is broken. The liquid flows out and solidifies at the surface where there is accommodation for it as the casting has contracted.

Inverse segregation is found to occur in tin bronzes. It is known by the name 'tin sweat'.

CRACKS

Cracks constitute a major defect in metals and alloys. They occur at the surface as well as just below the surface. Surface cracks are sometimes visible and sometimes not visible to the naked eye. The use of a small magnification is necessary to reveal the micro or hairline cracks. Cracks in general, and sub-surface cracks in particular, can be classified as intercrystalline and transcrystalline. Intercrystalline cracks occur along the grain boundaries of the metal, whereas the transcrystalline cracks develop through the grains themselves.

The following are some of the types of cracks:

(i) Quench or hardening cracks

(ii) Temper cracks

(iii) Shrinkage cracks

(iv) Hot tears

(v) Grinding cracks

(vi) Forging cracks

(vii) Fatigue cracks

Quench Cracks

These are induced in the hardened steel under severe quenching conditions. The surface of the austenitised steel quickly attains the temperature of the quenching medium, whereas the core remains at a higher temperature. The cooled surface contracts and exerts a stress on the interior. Besides, the transformation of the chilled surface layers to martensite also presents complex stress conditions. Consequently the steel either warps or cracks.

To remedy the quench cracking one or more of the following precautions are recommended:

(a) Use of a less severe quench.

(b) Austempering or martempering as is found suitable, instead of direct hardening.

Temper Cracks

The unequal strains set up in the steel during the process of quenching may not always manifest themselves in the form of quench cracks. In such cases, the cracks are found to develop during the tempering process.

Avoidance of the rapid heating or high tempering temperature lessens the possibility of the occurrence of temper cracks.

Shrinkage Cracks

The reason for shrinkage cracking is the same as that for the quench cracks except the complex phase transformation phenomenon. These are best avoided by avoiding faster cooling rates (Fig. 18.8).

Fig. 18.8 Shrinkage crack.

(*Courtesy* : Indian Copper Information Centre, Calcutta).

Hot Tears

This is a casting defect arising out of improper mould design or pouring technique.

Grinding Cracks

When a smooth (damaged) abrasive wheel is used in grinding, the operator tends to employ higher pressures. Consequently, the surface of the material is over-heated. This localized heating sets uneven expansion conditions and cracking results (Fig. 18.9). These cracks are of varying sizes and very closely resemble the thermal cracks. They occur in groups and are roughly parallel to one another. When a cross-section is taken, it is revealed that the crack is very sharp at its root and widens towards the surface.

Fig. 18.9 Grinding cracks on a tool.
(*Courtesy* : Magnaflux Corporation, Chicago, Illinois, U.S.A.)

Grinding is a finishing operation in the engineering industry. Therefore these cracks are very dangerous from the point of view of fatigue strength.

Proper selection of the grinding wheel, avoidance of worn-out wheels and better control of the pressure of grinding are recommended to minimize the grinding cracks.

Forging Cracks

Cracks appear on the object being forged when:

(i) the temperature of the stock is low,

(ii) high forging pressures are used,

(iii) the forging is done too slowly, or

(iv) the forging stock contains non-metallic inclusions.

These cracks are not governed by the grain structure and appear in any direction. They can however be avoided by remedying the above four causes.

Fatigue Cracks

These develop during the usage of the material. The cracks start from the roots of the notches which are first to form. The notches are formed due to many reasons such as:

(i) Accidental blows received by the metal during careless handling and usage lead to scratches, abrasion and nicking on the surface;

(ii) De-carburised surface conditions, inclusions, etc.;

(iii) Faulty design, leading to sudden changes in section;

(iv) Invisible quench cracks, grooves developed in machining and filing, forging laps, etc., may ultimately lead to notch formation.

Tables 18A and 18B give a comprehensive list (available from the literature) of the common defects, their appearance, reasons for their occurrence and the possible methods of remedy.

SALVAGE OF CASTINGS

Generally the castings made are of two types:

1. those which need very little or no machining, e.g., sewage covers, electric motor bodies, fly wheels, etc., and
2. those which need machining, e.g., auto engine block, head, etc.

Casting defects will manifest in many ways as described above. It is not always economical nor advisable to scrap every defective casting. Whenever a crack or blow hole is noticed (when the machining is done or otherwise) its location should be carefully studied. In castings where the strength factor is not involved, the cracks or holes can be filled up by weld metal.

The problem arises with precision castings and castings involving strength requirements. Iron and steel castings can be salvaged by filling in a paste made out of Araldite*, a patented product of M/s. Ciba Limited of Switzerland. The Araldite adhesive and hardener (supplied in separate containers) are mixed in recommended proportions and iron powder (less than 100 mesh size) is added to it and thoroughly mixed to the texture of a paste. This paste is thoroughly filled into the casting blow holes and they are allowed to set for 24 hours. Alternative methods of setting for shorter times at higher temperatures are recommended but the author, by his experience, feels that the room temperature setting is the best. Later, the castings can be machined and used. This method is successfully employed in M/s. General Motors Ltd., U.S.A., and in fact was recommended by them to M/s, Hindustan Motors Ltd., India, where the author has worked previously.

Of late, patented products like plastic steel are being marketed. It contains a synthetic plastic and steel in varying proportions as the manufacturers claim it. Its usefulness with respect to salvaging needs some more attention.

Table 18 A Casting defects.

Sl. No.	Defect	Appearance	Reasons for the occurrence of the defect	Suggested methods for elimination
1	**2**	**3**	**4**	**5**
1.	Pipe.	A conical depression at the top, or a hollow cavity in the interior of the casting (V -shaped).	The casting contracts considerably on cooling as the solidification commences at the surface of the mould.	By using a brick-lined-top on the mould, in which the metal remains molten for a prolonged period (feeder head). This liquid compensates for the shrinkage in the mould.
2.	Blow holes or entrapped gas.	Balloon-shaped gaseous cavities at or below the casting surface.	It is found that molten steel dissolves or occludes considerable quantities of gases. These gases are given out, very largely, during solidification but being unable to escape out, are liable to remain entrapped in the form of bubbles of very regular shapes.	By reducing the amount of gases formed by the mould reaction or improving its permeability, minimizing superheat and using moisture-free fluxes and clean and dry implements.
3.	Metal penetration.	A spongy mass of sand and metal which adheres tightly to the casting.	Metal may either penetrate into the sand because it is mechanically soft and porous or on account of chemical reaction between the metal and sand causing melting of the sand grains by the metal.	Harder ramming, improved sand flowability, use of a mould coating, and the use of special facing sand.

Table 18 A *Contd…*

Sl. No.	Defect	Appearance	Reasons for the occurrence of the defect	Suggested methods for elimination
1	**2**	**3**	**4**	**5**
5.	Core raise.	Irregular shape of the casting due to the floating of core (moves up close to the cope).	A weak core or one which is not properly supported with chapplets. An under-size core not held tightly by the core prints.	Raising the dry strength and properly supporting the core with chapplets.
6.	Core shift.	One side of the casting wall is thinner and the opposite one thicker than required.	The core is not securely held by chapplets or prints, or it was not centred when set.	Support the core properly by chapplets or proper prints. Correct centring is necessary while setting the core.
7.	Core wash.	Rough surface in the cavity.	Weak sand, if low in strength or with insufficient moisture.	By using a refractory coating on the core or strengthening it by increasing the binder or use of an arbour.
8.	Hot tears.	The casting may rupture or tear.	Too high a hot strength of the sand mixture. More percentage of inorganic binders may also lead to this defect.	Lowering hot strength of the core or by increasing the collapsibility of the core after solidification of metal.
9.	Veins.	Fin shaped protrusions in the cored areas.	A crack-shaped opening develops in the core as it becomes heated by the liquid metal. The crack is then filled with the molten metal.	Certain materials should be added to the core sands, (such as iron oxide) that flux the sand and make it tough at high temperature.
10.	Scabs.	Incrustations of sand and metal on the surface of the casting.	A portion of the mould surface spalling or flaking off because of the thermal expansion of sand.	Use of moulding sand of greater thermal stability. Softer ramming of the mould, reduction of dry strength, reduction of hot strength, use of combustibles or cushioning in the sand, better mixing, etc.

Table 18 A *Contd…*

Sl. No.	Defect	Appearance	Reasons for the occurrence of the defect	Suggested methods for elimination
1	2	3	4	5
12.	Fissures.	Rough, grainy looking masses attached directly to the cored surfaces of the castings.	The core sand has collapsed and been pushed away by the still molten metal.	Raising the hot strength of the core.
13.	Misruns or unfilled moulds.	Incomplete casting.	Low pouring temperature of metal. The metal solidifies before the mould is completely filled.	Employment of a higher pouring temperature (with more super heat).
14.	Fins.	Appearance of depression lines of the metal in the casting cavity.	Fins and projections on the core.	The fins and projections on the core should be cleaned off during the core finishing operations.
15.	Mould crack.	A fin on the casting.	Weak mould or a mould not properly supported, may crack. Metal runs into the crack and solidifies.	Stronger and properly supported mould.
16.	Cold shut.	Partial solidification of the stream before the mould is properly filled.	Too slow a pouring or too low a pouring temperature.	Faster pouring and more superheat to the liquid metal.
17.	Pin holes.	Numerous small gas cavities at or slightly below the surface.	Gases under pressure or excessive moisture and poor permeability of the mould.	Reducing the amount of gases formed by the mould. Improving its permeability or lowering the moisture content.

Table 18 A *Contd…*

Sl. No.	Defect	Appearance	Reasons for the occurrence of the defect	Suggested methods for elimination
1	2	3	4	5
19.	Drop.	Irregular casting.	Weak sand or rough mould handling. A portion of the cope drops into the drag.	Use of a sand of higher bond strength, moulding to greater hardness and better sand mixing.
20.	Shrinks.	The casting shows some enlargement in dimensions or is over weight. When this defect occurs, the casting is obviously bulging or over-size.	The sand may be too soft. The metallostatic pressure causes the mould cavity to enlarge.	Raising of hardnees of the mould using a sand of higher thermal stability (refractoriness).
21.	Segregation.	Different compositions at different portions of a casting.	Different constituents of the alloy possess various degrees of fusibility. Some remain in the liquid state after the rest have solidified. Slow cooling is generally the cause for segregation (normal segregation).	Rapid solidification. Normalizing treatment after the casting has solidified.
22.	Crush.	The defect shows on the casting as a depression where the mould wall is crushed in, towards the mould cavity.	Sand on the mould cavity is displaced by the cores. If a core is over-size, the core prints will be enlarged by the core when the mould is closed. This can flake the sand off the mould walls. Careless closing of the mould or poor flask guide pins may also cause this trouble.	Use of correct sized cores, closing the mould with care and with proper flask-guide pins.

Table 18 A *Contd…*

Sl. No.	Defect	Appearance	Reasons for the occurrence of the defect	Suggested methods for elimination
1	**2**	**3**	**4**	**5**
24.	Unsoundness or porosity.	Tiny pores throughout the casting.	Dissolved gases or high moisture or low permeability in the mould.	Decrease the moisture content, increase the permeability of the moulds. Use of mould coatings is also recommended.
25.	Inclusions.	Foreign materials in the casting.	Weak sand and dirt in the mould. Improper cleaning of the ladles, oxides in the metal.	Cleaning of the mould before casting, proper ramming of the mould and using clean ladles. Good control during the metal melting.
26.	Cold shot.	Globules of solid metals entrapped into the casting.	Formed by splashing during pouring, they are not dissolved later by the surrounding liquid metal.	By using proper pouring techniques. Bottom pouring is best. Otherwise a splash core should be used.
27.	Rough surface.	The surface of the casting will be rough.	Unfinished mould walls, coarse sand, drift in the mould, etc.	Using a fine grained sand and properly finishing the mould.
1	**2**	**3**	**4**	**5**
28.	Burning on.	Sand particles will firmly adhere to the casting surface.	The moulding sand has a low refractoriness. Incipient flusion occurs and the sand grains get stuck to the casting.	Use a sand of higher refractoriness or employ better mould dressings.
29.	Draw.	Saucer shaped depression on heavy sections, usually with rough edges.	Incorrect gating and feeding.	Use of risers to feed heavy sections.

*An epoxy resin

TABLE 18 B Fabrication defects.

Sl. No.	Defect	How the defect occurs	Reasons for the occurrence of the defect	Suggested method for rectification or avoidance
1	2	3	4	5
Rolling defects				
1.	Fins and over-fills.	Protrusions on the section.	When the section is too large for the pass it is entering or when proper allowance has not been made for lateral spreading during pass design.	Feeding correct size of material into the pass and (providing for lateral) spreading allowance in the pass design.
2.	Under-fills.	These are the areas in which the section is incompletely filled.	Formed by permitting the bar to be rolled slant-wise in certain directions.	Feeding adequate material into the pass.
3.	Slivers.	These are loose or torn segments of steel rolled into the surface of the bar.	The bar shearing against a guide or collar; incorrect entry into a closed pass.	Feeding adequate material into the pass.
4.	Lap.	Protrusion rolled into the surface of the product.	This may be due to the rolled over condition caused by a bar having been given a pass through the rolls after a sharp over-fill or fin has formed.	Feeding adequate material into the pass.
5.	Seam.	Crevice in the steel that has been closed but not welded.	Blow-holes and cracks in the original ingots.	By inspecting the feed-material, and correct selection.

Table 18 B *Contd...*

Sl. No.	Defect	How the defect occurs	Reasons for the occurrence of the defect	Suggested method for rectification or avoidance
1	2	3	4	5
Rolling defects				
6.	Fine cracks and roll marks.	Impressions in the product.	Over-heated mill rolls, cracking or spalling of the same.	By periodic checking of the rolls
7.	Scratches.	Long nicks or indentations in the product.	Surface or surfaces of the bar rubbing against sharp or pointed object such as guides in the mill.	Greater care in feeding.
8.	Rolled in oxide.	Scaly appearance.	Scale formed during the previous heating.	Scale formed during the heating of the steel should be completely removed, before feeding the stock into the rolls.
9.	Buckle and kink.	Corrugated or wrinkled surface.	Caused by either worn-out pinions on a roll stand or uneven cooling.	Careful cooling.
10.	Burned steel.	Rough area.	Incipient fusion and oxidation along grain boundaries.	Permanent defect and cannot be rectified. The material should be scrapped.
11.	Hook.	Sharp bend or curvature.	Improperly adjusted delivery guides or any obstruction which may halt the forward motion.	Proper adjustment of the guides.
12.	Camber.	Deviation of the side edge of a bar from a straight line.	Improper heating of the billet or uneven dimensions causing differential expansion.	Proper heating of the billet.

Table 18 B *Contd…*

Sl. No.	Defect	How the defect occurs	Reasons for the occurrence of the defect	Suggested method for rectification or avoidance
1	2	3	4	5
14.	Shear distortion.	Mashed or deformed end.	Defective or improperly adjusted shearing adjustment.	Use correct shear.
Extrusion Defects:				
1.	Extrusion defect.	An annular ring of oxide on a transverse section throughout the extrusion.	Inhomogeneous deformation in the direct extrusion of the billet, the centre of the billet moves faster than the periphery.	To carry on the extrusion operation only to a point where the surface oxide begins to enter the die and then discard the remainder of the billet.
Forging Defects:				
1.	Surface cracks.	Cracks.	Result of excessive working of the surface at too low a temperature or hot-shortness.	(i) Very high pressure should be avoided. (ii) The billet should be heated to the adequate temperature of forging.
2.	Cold shut.	Discontinuity in the product.	When two surfaces of a metal fold against each other without getting welded up.	(i) Adequate heating of the billet. (ii) Reducing sulphur concentration in the reheating furnace.

Index

Index

A

B

C

D

L

M

N

O

P

Q

R

S

T

U

V

W

X

Y

9 789352 300372

Printed at Repro India Ltd.